Principles of Radio Communication

Principles of Radio Communication

Edited by
Fraidoon Mazda
MPhil DFH CEng FIEE

With specialist contributions

Focal Press
An imprint of Butterworth-Heinemann
Linacre House, Jordan Hill, Oxford OX2 8DP
A division of Reed Educational and Professional Publishing Ltd

$\mathcal{R}$ A member of the Reed Elsevier plc group

OXFORD BOSTON JOHANNESBURG
NEW DELHI SINGAPORE MELBOURNE

First published 1996

British Library Cataloguing in Publication Data
Mazda, Fraidoon F
    Principles of Radio Communications
    I. Title
    621.382

ISBN 02405 1457 2

Library of Congress Cataloguing in Publication
Mazda, Fraidoon F.
    Principles of Radio Communications/Fraidoon Mazda
    p. cm.
    Includes bibliographical references and index.
    ISBN 02405 1457 2
    1. Telecommunications. I. Title
    TK5101.M37 1993   92-27846
    621.382–dc20   CIP
Printed and bound in Great Britain

Contents

Preface

From a relatively humble beginning radio communications has seen an explosive growth over the past few years. Much of this has been in mobile voice and data applications, and these are described in a companion volume to this book. The present book introduces the principles which are common to all forms of radio based communication systems.

The earth's ionosphere and troposphere provide an important media for the transmission of radio waves, and their characteristics are described in Chapter 1. The principles of radiowave propagation, for both terrestrial and satellite based systems, are then described in Chapter 2.

The radio spectrum is a scarce natural resource, and the considerations, both technical and regulatory, in its effective management are described in Chapter 3.

Chapter 4 covers the design and construction of antennas, which are one of the most fundamental components for radio based communication systems. Chapter 5 then describes microwave radio communication techniques, and their applications in urban and rural point to multipoint systems are further developed in Chapter 6.

Eight authors have contributed to this book, all specialists in their field, and the success of the book is largely due to their efforts. The book is also based on selected chapters which were first published in the much larger volume of the *Telecommunications Engineers' Reference Book*.

Fraidoon Mazda
Bishop's Stortford
April 1996

List of contributors

P A Bradley
BSc MSc CEng MIEE
Rutherford Appleton Laboratory
(Sections 1.1–1.6)

Jim Giacobazzi
TeleSciences Transmission
Systems
(Chapter 5)

Mark Holker
CEng FIEE MBIM
Hiltek Ltd.
(Chapter 2)

J A Lane
DSc CEng FIEE FInstP
Radio Communications Agency
(Sections 1.7–1.11)

Fraidoon Mazda
MPhil DFH CEng FIEE
Nortel Ltd
(Chapter 7)

S A Mohamed
BSc MSc PhD CEng MIEE
BT Laboratories
(Chapter 6)

Professor A D Olver
BSc PhD CEng FIEE FIEEE
Queen Mary & Westfield College
(Chapter 4)

David J Withers
CEng FIEE
Telecommunications Consultant
(Chapter 3)

1. Ionosphere and troposphere

1.1 The ionosphere

The ionosphere is an electrified region of the earth's atmosphere situated at heights of from about fifty kilometres to several thousand kilometres. It consists of ions and free electrons produced by the ionising influences of solar radiation and of incident energetic solar and cosmic particles.

The ionosphere is subject to marked geographic and temporal variations. It has a profound effect on the characteristics of radio waves propagated within or through it. By means of wave refraction, reflection or scattering it permits transmission over paths that would not otherwise be possible, but at the same time it screens some regions that could be illuminated in its absence (see Figure 1.1).

The ability of the ionosphere to refract, reflect or scatter rays depends on their frequency and elevation angle. Ionospheric refraction is reduced at the higher frequencies and for the higher elevation angles, so that provided the frequency is sufficiently great rays 1 in Figure 1.1 escape whereas rays 2 are reflected back to ground. Rays 3 escape because they traverse the ionosphere at latitudes where the electron density is low (the plasmapause trough (Muldrew, 1965)). Irregularities in the F-region are responsible for the direct backscattering of rays 4. The low-elevation rays 5 are reflected to ground because of the increased ionisation at the higher latitudes. Note that for this ionosphere and frequency there are two ground zones which cannot be illuminated.

The ionosphere is of considerable importance in the engineering of radio communication systems because:

1.   It provides the means of establishing various communication paths, calling for system-design criteria based on a knowledge of ionospheric morphology.

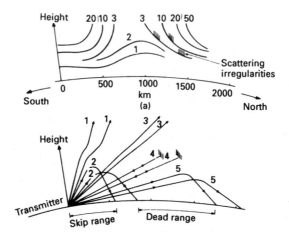

Figure 1.1 High frequency propagation paths via the ionosphere at high latitudes: (a) sample distribution of electron density (arbitrary units) in northern hemisphere high-latitude ionosphere (adapted from Buchau, 1972); (b) ray paths for signals of constant frequency launched with different elevation angles

2. It requires specific engineering technologies to derive experimental probing facilities to assess its characteristics, both for communication-systems planning and management, and for scientific investigations.
3. It permits the remote monitoring by sophisticated techniques of certain distant natural and man-made phenomena occurring on the ground, in the air and in space.

1.2 Formation of the ionosphere and its morphology

There is widespread interest in the characteristics of the ionosphere by scientists all over the world. Several excellent general survey books have been published describing the principal known features (Davies, 1990; Ratcliffe, 1970) and other more specialised books concerned with aeronomy, and including the ionosphere and magnetosphere, are of great value to the research worker (Rishbeth and Garriott, 1969; Davies, 1969; Ratcliffe, 1972). Several journals in the English language are devoted entirely, or to a major extent, to papers describing investigations into the state of the ionosphere and of radio propagation in the ionosphere.

The formation of the ionosphere is a complicated process involving the ionising influences of solar radiation and solar and cosmic particles on an atmosphere of complex structure. The rates of ion and free-electron production depend on the flux density of the incident radiation of particles, as well as on the ionisation efficiency, which is a function of the ionising wavelength (or particle energies) and the chemical composition of the atmosphere. There are two heights where electron production by the ionisation of molecular nitrogen and atomic and molecular oxygen is a maximum. One occurs at about 100km and is due to incident X-rays with wavelengths less than about 10nm and to ultraviolet radiation with wavelengths near 100nm; the other is at about 170km and is produced by radiation of wavelengths 20- 80nm.

Countering this production, the free electrons tend to recombine with the positive ions and to attach themselves to neutral molecules to form negative ions. Electrons can also leave a given volume by diffusion or by drifting away under the influences of temperature and pressure gradients, gravitational forces or electric fields set up by the movement of other ionisation. The electron density at a given height is governed by the so-called Continuity equation in terms of the balance between the effects of production and loss.

Night-time electron densities are generally lower than in the daytime because the rates of production are reduced. Figure 1.2 gives examples of a night-time and a daytime height distribution of electron density. The ionisation is continuous over a wide height range, but there are certain height regions with particular characteristics, and these are known, following E.V. Appleton, by the letters D, E and F. The E-region is the most regular ionospheric region, exhibiting a systematic dependence of maximum electron density on solar-zenith

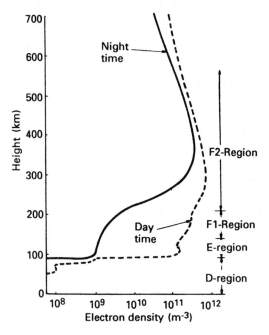

Figure 1.2   Sample night-time and daytime height distributions of electron density at mid-latitudes in summer

angle, leading to predictable diurnal, seasonal and geographical variations. There is also a predictable dependence of its electron density on the changes in solar radiation which accompany the long-term fluctuations in the state of the sun. Maximum E-region electron density is approximately proportional to sunspot number, which varies over a cycle of roughly 11 years.

In the daytime the F-region splits into two, with the lower part known as the F1-region and the upper part as the F2-region. This splitting arises because the principal loss mechanism is an ion- atom interchange process followed by dissociative recombination, the former process controlling the loss rates in the F2-region and the latter in the F1-region. Although maximum production is in the F1-region, maximum electron density results in the F2-region, where the loss rates are lower. The maximum electron density of the F1-region closely follows that of the E-region, but there are significant and less predictable changes in its height.

The maximum electron density and height of the F2-region are subject to large changes which have important consequences to radio wave propagation. Some of these changes are systematic but there are also major day-to-day variations. The F2-region is controlled mainly by ionisation transport to different heights along the lines of force of the earth's magnetic field under the influence of thermospheric winds at high and middle latitudes (Rishbeth, 1972) and by electric fields at low latitudes (Duncan, 1960).

These effects, taken in conjunction with the known variations in atmospheric composition, can largely explain characteristics of the F2-region which have in the past been regarded as anomalous by comparison with the E-region, namely, diurnal changes in the maximum of electron density in polar regions in the seasons of complete darkness, maximum electron densities at some middle-latitude locations at times displaced a few hours from local noon with greater electron density in the winter than the summer, and at low latitudes longitude variations linked more to the magnetic equator than to the geographic equator, with a minimum of electron density at the magnetic equator and maxima to the north and south where the magnetic dip is about 30°.

At all latitudes electron densities in the F2-region, like those in the E- and F1-regions, increase with increase of sunspot number. The electron densities at heights above the maximum of the F2-region are controlled mainly by diffusion processes.

The D-region shows great variability and fine structure, and is the least well understood part of the ionosphere. The only ionising radiations that can penetrate the upper regions and contribute to its production are hard X-rays with wavelengths less than about 2nm and Lyman-α radiation at 121.6nm. Chemical reactions responsible for its formation principally involve nitric oxide and other minor atmospheric constituents.

The D-region is mainly responsible for the absorption of radio waves because of the high electron-collision frequencies at such altitudes (see below). While the electron densities in the upper part of the D-region appear linked to those in the E-region, leading to systematic latitudinal, temporal and solar-cycle variations in absorption, there are also appreciable irregular day-to-day absorption changes. At middle latitudes anomalously high absorption is experienced on some days in the winter. This is related to warmings of the stratosphere and is probably associated with changes in D-region composition. In the lower D-region at heights below about 70km the ionisation is produced principally by energetic cosmic rays, uniformly incident at all times of day.

Since the free electrons thereby generated tend to collide and become attached to molecules to form negative ions by night, but are detached by solar radiation in the daytime, the lower D-region ionisation, like that in the upper D-region, is much greater by day than night. In contrast, however, electron densities in the lower D-region, being related to the incidence of cosmic rays, are reduced with increase in the number of sunspots.

Additional D-region ionisation is produced at high latitudes by incoming particles, directed along the lines of force of the earth's magnetic field. Energetic electrons, probably originating from the sun, produce characteristic auroral absorption events over a narrow band of latitudes about 10° wide, associated with the visual auroral regions (Hartz, 1968).

From time to time disturbances occur on the sun known as solar flares. These are regions of intense light, accompanied by increases in the solar far ultraviolet and soft X-ray radiation. Solar flares are most common at times of high sunspot number. The excess radiation leads to sudden ionospheric disturbances (SIDs), which are rapid and large increases in ionospheric absorption occurring simultaneously over the whole sunlit hemisphere. These persist from a few minutes to several hours giving the phenomena of short-wave fadeouts (SWFs), first explained by Dellinger.

Accompanied by solar flares are eruptions from the sun of energetic protons and electrons. These travel as a column of plasma, and depending on the position of the flare on the sun's disc and on the trajectory of the earth, they sometimes impinge on the ionosphere. Then, the protons, which are delayed in transit from fifteen minutes to several hours, produce a major enhancement of the D-region ionisation in polar regions that can persist for several days. This gives the phenomenon of polar cap absorption (PCA) with complete suppression of HF signals over the whole of both polar regions (Bailey, 1964). Slower particles, with transit times of 20-40 hours, produce ionospheric storms. These storms, which result principally from movements in ionisation, take the form of depressions in the maximum electron density of the F2-region (Matsushita, 1967). They can last for several days at a time with effects which are progressively different in detail at different latitudes. Since the sun rotates with a period of about 28 days, sweeping out a column of particles into space when it is disturbed, there is a tendency for ionospheric storms to recur after this time interval.

Additional ionisation is sometimes found in thin layers, 2km or less thick, embedded in the E-region at heights between 90 and 120km. This has an irregular and patchy structure, a maximum electron density which is much greater than that of the normal E-region, and is known as sporadic-E (or Es) ionisation because of its intermittent occurrence.

It consists of patches up to 2000km in extent, composed of large numbers of individual irregularities each less than 1km in size. Sporadic-E tends to be opaque to the lower h.f. waves and partially reflecting at the higher frequencies. It results from a number of separate causes and may be classified into different types (Smith and Matsushita, 1962), each with characteristic occurrence and other statistics. In temperate latitudes sporadic-E arises principally from wind shear, close to the magnetic equator it is produced by plasma instabilities and at high latitudes it is mainly due to incident energetic particles. It is most common at low latitudes where it is essentially a daytime phenomenon.

Irregularities also develop in the D-region due to turbulence and wind shears and other irregularities are produced in the F-region. The F-region irregularities can exist simultaneously over a wide range of heights, either below or above the height of maximum electron density, and are referred to as spread-F irregularities. They are found at all latitudes but are particularly common at low latitudes in the evenings where their occurrence is related to rapid changes in the height of the F-region (Cohen and Bowles, 1961).

1.3 Ionospheric effects on radio signals

A radio wave is specified in terms of five parameters: its amplitude, phase, direction of propagation, polarisation and frequency. The principal effects of the ionosphere in modifying these parameters are considered as follows.

1.3.1 Refraction

The change in direction of propagation resulting from the traverse of a thin slab of constant ionisation is given approximately by Bouger's law in terms of the refractive index and the angle of incidence. A more exact specification including the effects of the earth's magnetic field is given by the Haselgrove equation solution (Haselgrove, 1954). The refractive index is determined from the Appleton-Hartree equations of the magnetoionic theory (Ratcliffe, 1959; Budden, 1961), as a function of the electron density and electron-collision frequency, together with the strength and direction of the earth's magnetic field, the wave direction and the wave frequency. The dependence on frequency leads to wave dispersion of modulated signals. Since the ionosphere is a doubly refracting medium it can transmit two waves with different polarisations (see below).

The refractive indices appropriate to the two waves differ. Refraction is reduced at the greater wave frequencies, and at VHF and higher frequencies it is given approximately as a function of the ratio of the wave and plasma frequencies, where the plasma frequency is defined in terms of an universal constant and the square root of the electron density (Ratcliffe, 1959).

Table 1.1 lists the magnitude of the refraction and of other propagation parameters for signals at a frequency of 100MHz which traverse the whole ionosphere.

1.3.2 Change in phase-path length

The phase-path length is given approximately as the integral of the refractive index with respect to the ray-path length. Ignoring spatial

Table 1.1   Effect of one-way traverse of typical mid-latitude ionosphere at 100MHz on signals with elevation angle above 60 degrees (ITU-R, 1995a)

Effect	Day	Night	Frequency dependence (f)
Total electron content	5×10^{13} per cm^2	5×10^{12} per cm^2	
Faraday rotation	15 rotations	1.5 rotations	f^{-2}
Group delay	12.5μs	1.2μs	f^{-2}
Change in phase-path length	5.2km	0.5km	f^{-2}
Phase change	7500 radians	750 radians	f^{-2}
Phase stability (peak-to-peak)	±150 radians	±15 radians	f^{-1}
Frequency stability (r.m.s.)	±0.04Hz	±0.004Hz	f^{-1}
Absorption	0.1dB	0.01dB	f^{-2}
Refraction	≤1	Negligible	f^{-2}

gradients, the change in phase-path length introduced by passage through the ionosphere to the ground of signals at VHF and higher frequencies from a spacecraft is proportional to the total electron content. This is the number of electrons in a vertical column of unit cross-section.

1.3.3 Group delay

The group and phase velocities of a wave differ because the ionosphere is a dispersive medium. The ionosphere reduces the group velocity and introduces a group delay which for transionospheric signals at VHF and higher frequencies, like the phase-path change, is proportional to the total electron content.

1.3.4 Polarisation

Radio waves that propagate in the ionosphere are called characteristic waves. There are always two characteristic waves known as the ordinary wave and the extraordinary wave; under certain restricted conditions a third wave known as the Z-wave can also exist (Ratcliffe, 1959). In general the ordinary and extraordinary waves are elliptically polarised. The polarisation ellipses have the same axial ratio, orientations in space that are related such that under many conditions they are approximately orthogonal, and electric vectors which rotate in opposite directions (Ratcliffe, 1959).

The polarisation ellipses are less elongated the greater the wave frequency. Any wave launched into the ionosphere is split into characteristic ordinary and extraordinary wave components of appropriate power. At MF and above these components may be regarded as travelling independently through the ionosphere with polarisations which remain related, but continuously change to match the changing propagation path and associated ionospheric conditions. The phase paths of the ordinary and extraordinary wave components differ, so that in the case of transionospheric signals when the components have comparable amplitudes, the plane of polarisation of their resultant slowly rotates. This effect is known as Faraday rotation.

1.3.5 Absorption

Absorption arises from inelastic collisions between the free electrons, oscillating under the influence of the incident radio wave, and the neutral and ionised constituents of the atmosphere. The absorption experienced in a thin slab of ionosphere is given by the Appleton-Hartree equations (Ratcliffe, 1959) and under many conditions is proportional to the product of electron density and collision frequency, inversely proportional to the refractive index and inversely proportional to the square of the wave frequency. The absorption is referred to as non-deviative or deviative depending on whether or not it occurs where the refractive index is close to unity. Normal absorption is principally a daytime phenomenon. At frequencies below 5MHz it is sometimes so great as to completely suppress effective propagation. The absorptions of the ordinary and extraordinary waves differ, and in the range 1.5-10MHz the extraordinary wave absorption is significantly greater.

1.3.6 Amplitude fading

If the ionosphere were unchanging the signal amplitude over a fixed path would be constant. In practice, however, fading arises as a consequence of variations in propagation path, brought about by movements or fluctuations in ionisation. The principal causes of fading are:

1. Variations in absorption.
2. Movements of irregularities producing focusing and defocusing.
3. Changes of path length among component signals propagated via multiple paths.
4. Changes of polarisation, such as for example due to Faraday rotation.

These various causes lead to different depths of fading and a range of fading rates. The slowest fades are usually those due to absorption changes which have a period of about 10 minutes. The deepest and most rapid fading occurs from the beating between two signal components of comparable amplitude propagated along different paths. A regularly reflected signal together with a signal scattered from spread-F irregularities can give rise to so-called flutter fading, with fading rates of about 10Hz. A good general survey of fading effects, including a discussion of fading statistics, has been produced (CCIR, 1990). On operational communication circuits fading may be combated by space diversity or polarisation-diversity receiving systems and by the simultaneous use of multiple-frequency transmissions (frequency diversity).

1.3.7 Frequency deviations

Amplitude fading is accompanied by associated fluctuations in group path and phase path, giving rise to time and frequency-dispersed signals. When either the transmitter or receiver is moving, or there are systematic ionospheric movements, the received signal is also Doppler-frequency shifted.

Signals propagated simultaneously via different ionospheric paths are usually received with differing frequency shifts. Frequency shifts for reflections from the regular layers are usually less than 1Hz, but shifts of up to 20-30Hz have been reported for scatter-mode signals at low latitudes (Nielson, 1968).

1.3.8 Reflection, scattering and ducting

The combined effect of refraction through a number of successive slabs of ionisation can lead to ray reflection. This may take place over a narrow height range as at LF or rays may be refracted over an appreciable distance in the ionosphere as at HF Weak incoherent scattering of energy occurs from random thermal fluctuations in electron density, and more efficient aspect-sensitive scattering from ionospheric irregularities gives rise to direct backscattered and forward-scatter signals.

Ducting of signals to great distances can take place at heights of reduced ionisation between the E- and F-regions, leading in some cases to round-the-world echoes (Fenwick and Villard, 1963). Ducting can also occur within regions of field-aligned irregularities above the maximum of the F-region.

1.3.9 Scintillation

Ionospheric irregularities act as a phase-changing screen on transionospheric signals from sources such as earth satellites or radiostars. This screen gives rise to diffraction effects with amplitude, phase and angle-of-arrival scintillations (Ratcliffe, 1956).

1.4 Communication and monitoring

Ionospheric propagation is exploited for a wide range of purposes, the choice of system and the operating frequency being largely determined by the type and quantities of data to be transmitted, the path length and its geographical position.

1.4.1 Communication systems

Radio communication at very low frequencies (VLF) is limited by the available bandwidth, but since ionospheric attenuation is very low, near world-wide coverage can be achieved. Unfortunately the radiation of energy is difficult at such frequencies and complex transmitting antenna systems, coupled with large transmitter powers, are needed to overcome the high received background noise from atmospherics, the electromagnetic radiation produced by lightning discharges.

Despite the expanding introduction of alternate satellite links such as NAVSTAR and GPS, because of the stability of propagation, VLF systems continue to be used for the transmission of standard time signals and for CW navigation systems which rely on direction-finding techniques, or on phase comparisons between spaced transmissions as in the Omega system (10-14kHz) (Pierce, 1965). At low frequencies (LF) increased propagation losses limit area coverage, but simpler antenna systems are adequate and lower transmitter powers can be employed because of the reduced atmospheric noise. Low frequency systems are used for communication by on-off keying and frequency-shift keying.

Propagation conditions are more stable than at higher frequencies because the ionosphere is less deeply penetrated. Low frequency signals involving ionospheric propagation are also used for communication with submarines below the surface of the sea, with receivers below the ground and with space vehicles not within line-of-sight of the transmitter.

Other LF systems (Stringer, 1969) relying principally on the ground wave, which are sometimes detrimentally influenced by the sky wave at night, include the Decca CW navigation system (70-

130kHz), the Loran C pulse navigation system (100kHz) and long-wave broadcasting.

At medium frequencies (MF) daytime absorption is so high as to completely suppress the sky wave. Some use is made of the sky wave at night-time for broadcasting, but generally medium frequencies are employed for ground-wave services.

Despite the advent of reliable multichannel satellite and cable systems, high frequencies continue to be used predominantly for broadcasting, fixed and mobile point-to-point communications, via the ionosphere; there are still tens of thousands of such circuits.

Very high frequency (VHF) communication relying on ionos-pheric scatter propagation between ground-based terminals is possible.

Two-way error-correcting systems with scattering from intermit-tent meteor trains can be used at frequencies of 30-40MHz over ranges of 500-1500km (Sugar, 1964). Bursts of high-speed data of about one second duration with duty cycles of the order of 5% can be achieved, using transmitter powers of about 1kW.

Meteor-burst systems find favour in certain military applications because they are difficult to intercept, since the scattering is usually confined to 5-10 degrees from the great-circle path. Forward-scatter communication systems at frequencies of 30-60MHz, also operating over ranges of about 1000km, rely on coherent scattering from field-aligned irregularities in the D-region at a height of about 85km (Bailey et. al., 1955). They are used principally at low and high latitudes. Signal intensities are somewhat variable, depending on the incidence of irregularities.

During magnetically disturbed conditions signals are enhanced at high latitudes, but are little affected at low latitudes. Directional transmitting and receiving antennas with intersecting beams are re-quired.

Particular attention has to be paid to avoiding interference from signal components scattered from sporadic-E ionisation or irregu-larities in the F-region. Special frequency-modulation techniques involving time division multiplex are used to combat Doppler effects. Systems with 16 channels, automatic error correction and operating at 100 words per minute, now exist.

1.4.2 Monitoring systems

High frequency (HF) signals propagated obliquely via the ionosphere and scattered at the ground back along the reverse path may be exploited to give information on the characteristics of the scattering region (Croft, 1972). Increased scatter results from mountains, from cities and from certain sea waves. Signals backscattered from the sea are enhanced when the signal wavelength is twice the component of the sea wavelength along the direction of incidence of the signal, since round-trip signals reflected from successive sea-wave crests then arrive in phase to give coherent addition.

The Doppler shift of backscatter returns from sea waves due to the sea motion usually exceeds that imposed on the received signals by the ionosphere, so that Doppler filtering enables the land- and sea-scattered signals to be examined separately (Blair et al., 1969). Doppler filtering can also resolve signals reflected or scattered back along ionosphere paths from aircraft, rockets, rocket trails or ships. This technique is exploited in military over-the-horizon (OHD) radar systems which involve huge steered antenna arrays for fan-beam surveillance (Anderson, 1986; Headrick and Skolnik, 1974).

Studies of the Doppler shift of stable frequency, HF, CW signals propagated via the ionosphere between ground-based terminals provide important information about infrasonic waves in the F-region originating from nuclear explosions (Baker and Davies, 1968), earthquakes (Davies and Baker, 1965), severe thunderstorms (Baker and Davies, 1969), and air currents in mountainous regions.

High-altitude nuclear explosions lead to other effects which may be detected by radio means (Pierce, 1965). An immediate wideband electromagnetic pulse is produced which can be monitored throughout the world at VLF, and HF. Enhanced D-region ionisation, lasting for several days over a wide geographical area, produces an identifiable change in the received phase of long-distance VLF ionospheric signals. Other more localised effects which can be detected include the generation of irregularities in the F-region.

Atmospherics may be monitored at VLF out to distances of several thousand kilometres and by recording simultaneously the arrival azimuths at spaced receivers the locations of thunderstorms may be determined and their movements tracked as an aid to meteorological warning services.

1.5 Ionospheric probing techniques

There are a wide variety of methods of sounding the state of the ionosphere involving single- and multiple-station ground-based equipments, rocket-borne and satellite probes (see Table 1.2). A comprehensive survey of the different techniques has been produced by a Working Group of the International Union of Radio Science (URSI) (Smith, 1970). Some of the techniques involve complex analysis procedures and require elaborate and expensive equipment and antenna systems (Figure 1.3), others need only a single radio receiver.

Figure 1.3 The EISCAT radar antenna. (Phtotgraph courtesy of Rutherford Appleton Laboratory)

Table 1.2 Principal ionospheric probing techniques

Height range	Technique	Parameters monitored	Site
Above 100km	Vertical-incidence sounding	Up to height of maximum ionisation-electron density	Ground
	Topside sounding	From height of satellite to height of maximum ionisation-electron density; electron and ion temperatures; ionic composition	Satellite
	Incoherent scatter	Up to few thousand km - electron density; electron temperature; ion temperature; ionic composition; collision frequencies; drifts of ions and electrons	Ground
	Faraday rotation and differential Doppler	Total electron content	Satellite-ground
	In-situ probes	Wide range of parameters	Satellite
	CW oblique incidence	Solar flare effects; irregularities; travelling disturbances; radio aurorae	Ground
	Pulse oblique incidence	Oblique modes by ground backscatter and oblique sounding; meteors; radio aurorae; irregularities and their drifts	Ground
	Whistlers	Out to few Earth radii-electron density; ion temperature; ionic composition	Ground or satellite
Below 100km	Vertical-incidence sounding	Absorption	Ground
	Riometer	Absorption	Ground
	CW and pulse oblique incidence	Electron density and collision frequency	Ground
	Wave fields	Electron density and collision frequency	Rocket-ground
	In situ probes	Electron density and collision frequency; ion density; composition of neutral atmosphere	Rocket
	Cross-modulation	Electron density and collision frequency	Ground
	Partial reflection	Electron density and collision frequency	Ground
	Lidar	Neutral air density; atmospheric aerosols; minor constituents	Ground

The swept-frequency ground-based ionosonde consisting of a co-located transmitter and receiver was developed for the earliest of ionospheric measurements and is still the most widely used probing instrument. The transmitter and receiver frequencies are swept synchronously over the range from about 0.5-1MHz to perhaps 20MHz depending on ionospheric conditions, and short pulses typically of duration 100μs with a repetition rate of 50/s are transmitted. Calibrated film or computer digitised records of the received echoes give the group path and its frequency dependence.

In practice these records require expert interpretation because:

1.   Multiple echoes occur, corresponding to more than one traverse between ionosphere and ground (so-called multiple-hop modes) or when partially reflecting sporadic-E or F-region irregularities are present.

2.   The ordinary and extraordinary waves are sometimes reflected from appreciably different heights.

3.   Oblique reflections occur when the ionosphere is tilted.

Internationally agreed procedures for scaling ionosonde records (ionograms) have been produced (Piggott and Rawer, 1972). Automatic scaling software has been developed for use with digital ionograms (Reinisch and Huang, 1983) and comparisons with manual scaling lead to periodic refinements.

Since reflection takes place from a height where the sounder frequency is equal to the ionospheric plasma frequency, and since the group path can be related to that height provided the electron densities at all lower heights are known, the data from a full frequency sweep can be used to give the true-height distribution of electron density in the E- and F-regions up to the height of maximum electron density of the F2-region. The conversion of group path to true height requires assumptions regarding missing data below the lowest height from which echoes are received and over regions where the electron density does not increase monotonically with height. The subject of true-height analysis is complex and has been considered by several research groups for many years (Beynon, 1967). An internationally agreed procedure is now available (Titheridge, 1985). Commercially manufactured pulse sounders use transmitter powers of about 1kW. Sounders with powers of around 100kW and a lower frequency limit of a few kHz have been operated successfully in areas free from MF broadcast interference, to study the night time E-region. This has a maximum plasma frequency of around 0.5MHz.

Pulse-compression systems (Coll and Storey, 1964) and CW chirp sounders (Fenwick and Barry, 1966) offer the possibility of improved signal/noise ratios and echo resolution. In the chirp sounder system, originally developed for use at oblique incidence, the transmitter and receiver frequencies are swept synchronously so that the finite echo transit time leads to a frequency modulation of the receiver IF signals. These signals are then spectrum-analysed to produce conventional ionograms. Receiver bandwidths of only a few tens of hertz are needed so that transmitter powers of a few watts are adequate. Other ionosondes have been produced and used operationally which record data digitally on magnetic tape (Piggott and Rawer, 1972). An iono-sonde has been successfully flown in an aircraft to investigate geographical changes in electron density at high latitudes (Whalen et al., 1971). Over 100 ground-based ionosondes throughout the world make regular soundings each hour of each day; data are published at monthly intervals (Rutherford, 1991).

Over several years beginning in 1962 swept-frequency ionosondes have been operated in satellites orbiting the earth at altitudes of around 1000km. These are known as topside sounders and they give the distributions of electron density from the satellite height down to the peak of the F2-region. They also yield other plasma-resonance information, together with data on electron and ion temperatures and ionic composition. Fixed-frequency topside sounders have been used to study the spatial characteristics of spread-F irregularities and other features with fine structure.

Many different monitoring probes have been mounted in satellites orbiting the earth at altitudes above 100km, to give direct measurements of a range of ionospheric characteristics. These include RF impedance, capacitance and upper-hybrid resonance probes for local electron density, modified Langmuir probes for electron temperature, retarding potential analysers and sampling mass spectrometers for ion density, quadrupole and monopole mass spectrometers for ion and neutral-gas analysis and retarding potential analysers for ion temperature measurements.

Ground measurements of satellite beacon signals permit studies of total-electron content, either from the differential Doppler frequency between two harmonically related HF/VHF signals (Garriott and Nichol, 1965) or from the Faraday rotation of a single VHF transmission (Ross, 1965). Beacons on geostationary satellites are valuable for investigations of temporal variations. The scintillation of satellite signals at VHF and UHF gives information on the incidence of ionospheric irregularities, their heights and sizes (Aarons, 1973).

A powerful tool for ionospheric investigations up to heights of several thousand kilometres is the vertical-incidence incoherent-scatter radar. The technique makes use of the very weak scattering from random thermal fluctuations in electron density which exist in a plasma in quasi-equilibrium.

Several important parameters of the plasma affect the scattering such that each of these can be determined separately. The power, frequency spectrum and polarisation of the scattered signals are measured and used to give the height distributions of electron density, electron temperature, ionic composition, ion-neutral atmosphere and ion-ion collision frequencies, and the mean plasma-drift velocity. Tristatic receiving systems enable the vertical and horizontal components of the mean plasma drift to be determined. Radars operate at frequencies of 40-1300MHz using either pulse or CW transmissions. Transmitter peak powers of the order of 1MW, complex antenna arrays (Figure 1.3), and sophisticated data processing procedures are needed. Ground clutter limits the lowest heights that can be investigated to around 100km.

Electron densities, ion temperatures and ionic composition out to several earth radii may be studied using naturally occurring whistlers originating in lightning discharges. These are dispersed audio-frequency trains of energy, ducted through the ionosphere and then propagated backwards and forwards along the earth's magnetic-field lines to conjugate points in the opposite hemisphere. Whistler dispersions may be observed either at the ground or in satellites (Helliwell, 1965).

Continuous wave and pulsed signals, transmitted and received at ground-based terminals, may be used in a variety of ways to study irregularities or fluctuations in ionisation. Cross-correlation analyses of the amplitudes on three spaced receivers, of pulsed signals of fixed frequency reflected from the ionosphere at near vertical incidence, give the direction and velocity of the horizontal component of drift (Mitra, 1949).

The heights, patch sizes and incidence of F-region irregularities responsible for oblique-path forward-scatter propagation at frequencies around 50MHz may be investigated by means of highly-directional antennas and from signal transit times (Cohen and Bowles, 1961). Measurements of the Doppler frequency variations of signals from stable CW transmitters may be used to study:

1. Ionisation enhancements in the E- and F-regions associated with solar flares.
2. Travelling ionospheric disturbances (Munro, 1950).
3. The frequency-dispersion component of the ionospheric channel- scattering function (Bello, 1965).

Sporadic-E and F-region irregularities associated with visual aurorae may be examined by pulsed-radar techniques over a wide range of frequencies from about 6MHz to 3000MHz. They may also be investigated using CW bistatic systems in which the transmitter and receiver are separated by several hundred kilometres. Since the irregularities are known to be elongated and aligned along the direction of the earth's magnetic field and since at the higher frequencies efficient scattering can also occur under restricted conditions, the scattering centres may readily be located.

Using low-power VHF beacon transmitters, this technique has proved very popular with radio amateurs. Pulsed meteor radars incorporating Doppler measurements indicate the properties and movements of meteor trains (Sugar, 1964).

Two other oblique-path techniques, giving information on the regular ionospheric regions, are high-frequency ground backscatter sounding and variable-frequency oblique sounding. The former uses a nearby transmitter and receiver, and record interpretation generally involves identifying the skip distance (see Figure 1.1) where the signal returns are enhanced because of ray convergence. It is important to use antennas with azimuthal beamwidths of only a few degrees to minimise the ground area illuminated. Long linear antenna arrays with beam slewing, and circularly-disposed banks of log-periodic antennas with monopulsing are used. Oblique-incidence sounders are adaptions of vertical-incidence ionosondes with the transmitter and receiver controlled from stable-synchronised sources. Atlases of characteristic records obtained from the two types of sounder under different ionospheric conditions have been produced. Mean models of the ionosphere over the sounding paths may be deduced by matching measured data with ray-tracing results (Croft, 1968).

Large upwards-pointing antenna arrays fed from very high power transmitters operating at VLF through to UHF are capable of leading to ionospheric modification at F-region heights with the excitation of plasma instabilities by non-linear wave-interaction processes. Measurements mainly at HF are the subject of ongoing research involving a variety of phenomena including the generation of plasma waves and artificial aurorae (ITU-R, 1995b).

So far, no mention has been made of the height region below about 100km. As already noted, the D-region is characterised by a complex structure and high collision frequencies which lead to large daytime absorption of HF and MF waves. This absorption may be measured using fixed-frequency vertical-incidence pulses (Appleton and Piggott, 1954), or by monitoring CW transmissions at ranges of 200-500km, where there is no ground-wave component and the dominant signals are reflected from the E-region by day and the sporadic-E layer by night. There is then little change in the ray paths from day to night so that, assuming night absorption can be neglected, daytime reductions in amplitude are a measure of the prevailing absorption.

Multifrequency absorption data give information on the height distributions of electron density (Beynon and Rangaswamy, 1969). Auroral absorption is often too great to be measured in such ways, but special instruments known as riometers can be used (Hargreaves, 1969). These operate at a frequency around 30MHz and record changes in the incident cosmic noise at the ground caused by ionospheric absorption.

D-region electron densities and collision frequencies may be inferred from oblique or vertical-path measurements of signal amplitude, phase, group-path delay and polarisation at frequencies of 10Hz to 100kHz, with atmospherics as the signal sources at the lower frequencies. Vertically radiated signals in the frequency range 1.5-6MHz suffer weak partial reflections from heights of 75-90km. Measurements of the reflection coefficients of both the ordinary and extraordinary waves, which can be of the order of 10^{-5}, enable electron density and collision-frequency data to be deduced (Belrose and Burke, 1964).

Pulsed signals with high transmitter antenna systems and very sensitive receivers are needed. As well as in-situ probes in rockets, there are a wide range of other schemes for determining electron density and collision frequency, involving the study of wave-fields radiated between the ground and a rocket. These use combinations of frequencies in the VLF–VHF range and include the measurement of differential-Doppler frequency, absorption, differential phase, propagation time and Faraday rotation.

Theory shows that signals propagated via the ionosphere can become cross-modulated by high-power interfering signals which heat the plasma electrons through which the wanted signals pass. This heating causes the electron-collision frequency, and therefore the amplitude of the wanted signal to fluctuate at the modulation frequency of the interfering transmitter.

Investigations of this phenomenon (known as the Luxembourg effect after the first identified interfering transmitter) usually employ vertically transmitted and received wanted pulses, modulated by a distant disturbing transmitter radiating synchronised pulses at half the repetition rate. Changes in signal amplitude and phase between successive pulses are measured, and by altering the relative phase of the two transmitters, the height at which the cross-modulation occurs can be varied. Such data enable the height distributions of electron density in the D-region to be determined (Fejer, 1955).

Using a laser radar (lidar), the intensity of the light back-scattered by the atmospheric constituents at heights above 50km gives the height distributions of neutral-air density and the temporal and spatial statistics of high-altitude atmospheric aerosols. Minor atmospheric constituents may be detected with tunable dye lasers from their atomic and molecular-resonance scattering.

1.6 Propagation prediction procedures

Long-term predictions based on monthly median ionospheric data are required for the circuit planning of VLF–HF ground-based systems. Estimates of ray path launch and arrival angles are needed for antenna design, and of the relationship between transmitter power and received field strength at a range of frequencies, so that the necessary size of transmitter and its frequency coverage can be determined.

Since there are appreciable day to day changes in the electron densities in the F2-region, in principle short-term predictions based on ionospheric probing measurements or on correlations with geophysical indices should be of great value for real-time frequency management. In practice, however, aside from the technical problems of devising schemes of adequate accuracy:

1.   Not all systems are frequency agile (e.g. broadcasting).
2.   Effective schemes may require two-way transmissions.
3.   Only assigned frequencies may be used.

An alternative to short-term predictions is real-time channel sounding; certain procedures involve a combination of the two techniques.

1.6.1 Long-term predictions

The first requirement of any long-term prediction is a model of the ionosphere. At VLF waves propagate between the earth and the lower boundary of the ionosphere at heights of 70km by day and 90km by night as if in a two-surface waveguide. Very low frequency field-strength predictions are based on a full-wave theory that includes diffraction and surface-wave propagation.

For paths beyond 1000km range only three or fewer waveguide modes need to be considered. A general equation gives field strength as a function of range, frequency, ground-electrical properties and the ionospheric reflection height and reflection coefficients (ITU-R, 1995c). Unfortunately the reflection coefficients vary in a complex way with electron density and collision frequency, the direction and strength of the earth's magnetic field, wave frequency and angle of incidence, so that in the absence of accurate D-region electron-density data, estimates are liable to appreciable error.

At LF propagation is more conveniently described by wave-hop theory in terms of component waves with different numbers of hops. As at VLF reflection occurs at the base of the ionosphere and the accuracy of the field-strength prediction is largely determined by uncertainties in ionospheric models and reflection coefficients. Medium-frequency signals penetrate the lower ionosphere and are usually reflected from heights of 85-100km, except over distances of less than 500km by night when reflection may be from the F-region. Large absorptions occur near the height of reflection and daytime signals are very weak. It is now realised that because of the uncertainties in ionospheric models, signal-strength predictions are best based on empirical equations fitted to measured signal-strength data for other oblique paths.

Prediction schemes for HF tend to be complicated because they must assess the active modes and elevation angles; these vary markedly with ionospheric conditions and transmitter frequency. Equations are available for the ray paths at oblique incidence through ionospheric models composed of separate segments of simple analytic form (Appleton and Beynon, 1947; Croft and Hoogasian, 1968). Combinations of parabolic, quasiparabolic, linear and quasilinear segments are typically employed (Bradley and Dudeney, 1973; Dyson and Bennett, 1988) with the segment parameters determined from numerical prediction maps of the vertical-incidence ionospheric characteristics, as given by data from the world network of ionosondes (ITU-R, 1995d; see Figure 1.4).

Calculations over a fixed path for a range of frequencies indicate the largest frequency (the basic MUF or maximum usable frequency) that propagates via a given mode. Assuming some statistical law for the day to day variability of the parameters of the model they also give the availability, which is the fraction of days that the mode can exist. Received signal strengths are then determined in terms of the trans-

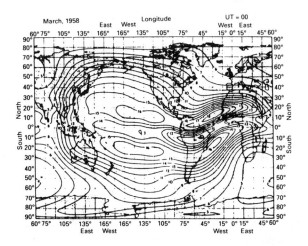

Figure 1.4 Predicted median, foF2 MHz for 00h, UT in March
1958. (Reproduced by permission of the Institute for
Telecommunication Sciences, Boulder, USA)

mitter power and a number of transmission loss and gain factors.
These include transmitting and receiving antenna gain, spatial attenuation, ray convergence gain, absorption, intermediate-path ground-
reflection losses and polarisation-coupling losses. Predictions may be
further extended by including estimates of the day to day variability
in signal intensity.

Calculations are prohibitively lengthy without computing aids and
a number of computerised prediction schemes have been produced
(ITU-R, 1995e; Teters et al., 1983). Nowadays such procedures are
available for microcomputer evaluation. By means of estimates of
background noise intensity, and from the known required signal/noise
ratio, the mode reliability may also be determined. This is the fraction
of the days that the signals are received with adequate strength. For

some systems involving fast data transmission, predictions of the probability of multipath, with two or more modes of specified comparable amplitude with propagation delays differing by less than some defined limit, are also useful and can be made.

1.6.2 Short-term predictions and real-time channel sounding

Some limited success has been achieved in the short-term prediction of the ionospheric characteristics used to give the parameters of the ionospheric models needed for HF performance assessment. Schemes are based either on spatial or temporal extrapolation of near real-time data or on correlations with magnetic activity indices. Regression statistics have been produced for the change in the maximum plasma frequency of the F2-region (foF2) with local magnetic activity index K, and other work is concerned with producing joint correlations with K and with solar flux.

In principle at HF the most reliable, although costly, way of ensuring satisfactory propagation over a given path and of optimising the choice of transmission frequency involves using an oblique-incidence sounder over the actual path; in practice, however, sounder systems are difficult to deploy operationally, require expert interpretation of their data, lead to appreciable spectrum pollution, and give much redundant information. Some schemes involve low-power channel monitoring of the phase-path stability on each authorised frequency, to ensure that at all times the best available is used. Real-time sounding on one path can aid performance predictions for another. Examples include ray tracing through mean ionospheric models simulated from measured backscatter or oblique-incidence soundings.

Many engineers operating established radio circuits prefer, for frequency management, to rely on past experience, rather than to use predictions. This is not so readily possible for mobile applications. Real-time sounding schemes involving ground transmissions on a range of frequencies to an aircraft, but only single-frequency transmission in the reverse direction, have proved successful (Stevens, 1968).

1.7 The troposphere

The influence of the lower atmosphere, or troposphere, on the propagation of radio waves is important, in several respects. At all frequencies above about 30MHz refraction and scattering caused by local changes in atmospheric structure become significant, especially in propagation beyond the normal horizon. In addition, at frequencies above about 5GHz, absorption in oxygen and water vapour in the atmosphere is important at certain frequencies corresponding to molecular absorption lines. An understanding of the basic characteristics of these effects is thus essential in the planning of very high frequency communication systems.

he main features of tropospheric propagation are summarised from a practical point of view as follows. There are two general problems: firstly, the influence of the troposphere on the reliability of a communication link. Here attention is concentrated on the weak signals which can be received for a large percentage of the time, say 99.99%. Secondly, it is necessary to consider the problem of interference caused by abnormal propagation and unusually strong, unwanted signals of the same frequency as the wanted transmission.

In both these aspects of propagation, the radio refractive index of the troposphere plays a dominant role. This parameter depends on the pressure, temperature and humidity of the atmosphere. Its vertical gradient and local fluctuations about the mean value determine the mode of propagation in many important practical situations. Hence the interest in the subject of radio meteorology, which seeks to relate tropospheric structure and radio-wave propagation. In most ground-to-ground systems the height range 0-2km above the earth's surface is the important region, but in some aspects of earth-space transmission, the meteorological structure at greater heights is also significant.

1.7.1 Historical background

Although some experiments on ultra-short-wave techniques were carried out by Hertz and others more than 90 years ago, it was only after about 1930 that any systematic investigations of tropospheric propagation commenced. For a long time it was widely believed that at frequencies above about 30MHz transmission beyond the geometric horizon would be impossible. However, this view was disputed by Marconi as early as 1932. He demonstrated that, even with relatively low transmitter powers, reception over distances several times the optical range was possible.

Nevertheless, theoreticians continued for several years to concentrate on studies of diffraction of ultra-short waves around the earth's surface. However, their results were found to over-estimate the rate of attenuation beyond the horizon. To correct for this disparity, the effect of refraction was allowed for by assuming a process of diffraction around an earth with an effective radius of 4/3 times the actual value. In addition, some experimental work began on the effect of irregular terrain and the diffraction caused by buildings and other obstacles.

However, it was only with the development of centimetric radar in the early years of World War II that the limitations of earlier concepts of tropospheric propagation were widely recognised. For several years attention was concentrated on the role of unusually strong refraction in the surface layers, especially over water, and the phenomenon of trapped propagation in a duct. It was shown experimentally and theoretically that in this mode the rate of attenuation beyond the horizon was relatively small. Furthermore, for a given height of duct or surface layer having a very large, negative, vertical gradient of refractive index, there was a critical wavelength above which trapping did not occur; a situation analogous to that in waveguide transmission.

Again however it became apparent that further work was required to explain experimental observations. The increasing use of VHF, and later UHF, for television and radio communication emphasised the need for a more comprehensive approach to beyond-the-horizon transmission. The importance of refractive-index variations at heights of the order of a kilometre began to be recognised and studies of the correlation between the height variation of refractive index and field strength began in several laboratories.

With the development of more powerful transmitters and antennae of very high gain it proved possible to establish communication well beyond the horizon even in a 'well-mixed' atmosphere with no surface ducts or large irregularities in the height variation of refractive index.

To explain this result, the concept of tropospheric scatter was proposed. The trans-horizon field was assumed to be due to incoherent scattering from the random, irregular fluctuations in refractive index produced and maintained by turbulent motion. This procedure has dominated much of the experimental and theoretical work since 1950 and it certainly explains some characteristics of troposphere propagation. However, it is inadequate in several respects. It is now known that some degree of stratification in the troposphere is more frequent than was hitherto assumed.

The possibility of reflection from a relatively small number of layers or sheets of large vertical gradient must be considered, especially at VHF At UHF and SHF strong-scattering from a 'patchy' atmosphere, with local regions of large variance in refractive index filling only a fraction of the common volume of the antenna beams, is probably the mechanism which exists for much of the time.

The increasing emphasis on microwaves for terrestrial and space systems has recently focused attention on the effects of precipitation on tropospheric propagation. While absorption in atmospheric gases is not a serious practical problem below 40GHz, the attenuation in rain and wet snow can impair the performance of links at frequencies of about 10GHz and above. Moreover scattering from precipitation may prove to be a significant factor in causing interference between space and terrestrial systems sharing the same frequency.

The importance of interference-free sites for earth stations in satellite links has also stimulated work on the shielding effect of hills and mountains. In addition, the use of large antennae of high gain in space systems requires a knowledge of refraction effects (especially at low angles of elevation), phase variations over the wavefront, and the associated effects of scintillation fading and gain degradation. Particularly at higher microwave frequencies, thermal noise radiated by absorbing regions of the troposphere (rain, clouds, etc.) may be significant in space communication. Much of the current research is therefore being directed towards a better understanding of the spatial structure of precipitation.

In addition, there has been a recent revival of interest in the effects of terrain (hills, buildings, etc.) at VHF, UHF and SHF, especially in relation to the increasing requirements of the mobile services.

1.8 Survey of propagation modes in the troposphere

Figure 1.5 illustrates qualitatively the variation of received power with distance in a homogeneous atmosphere at frequencies above about 30MHz. For antenna heights of a wavelength or more, the propagation mode on the free-space range is a space wave consisting of a direct and a ground-reflected ray. For small grazing angles the reflected wave has a phase change of nearly 180° at the earth's surface, but imperfect reflection reduces the amplitude below that of

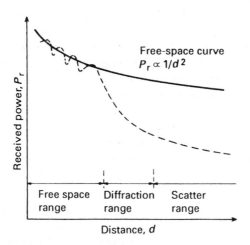

Figure 1.5   Tropospheric attenuation as a function of distance in a homogeneous atmosphere. Direct and ground reflected rays interfere in the free space range; obstacle diffraction effects predominate in the diffraction range; and refractive-index variations are important in the scatter range

the direct ray. As the path length increases, the signal strength exhibits successive maxima and minima. The most distant maximum will occur where the path difference is $\lambda/2$ where λ is the wavelength.

The range over which the space-wave mode is dominant can be determined geometrically allowing for refraction effects. For this purpose we can assume that the refractive index, n, decreases linearly by about 40 parts in 10^6 (i.e. 40 N units) in the first kilometre. This is the equivalent to increasing the actual radius of the earth by a factor of 4/3 and drawing the ray paths as straight lines. The horizon distance d, from an antenna at height h above an earth of effective radius a is given by Equation 1.1.

$$d = (2ah)^{\frac{1}{2}} \tag{1.1}$$

For two antennae 100m above ground the total range is about 82km, 15% above the geometric value.

Beyond the free-space range, diffraction around the earth's surface and its major irregularities in terrain is the dominant mode, with field strengths decreasing with increasing frequency and being typically of the order of 40dB below the free space value at 100km at VHF for practical antenna heights. As the distance increases, the effect of reflection or scattering from the troposphere increases and the rate of attenuation with distance decreases. In an actual inhomogeneous atmosphere the height-variation of n is the dominant factor in the scatter zone as illustrated in Figure 1.6. However, in practice the situation is rarely as simple as that indicated by these simple models.

At frequencies above about 40GHz, absorption in atmospheric gases becomes increasingly important. This factor may determine system design and the extent to which co-channel sharing is possible; for example, between terrestrial and space communication services. There are strong absorption lines due to oxygen at 60 and 119GHz, with values of attenuation, at sea level, of the order of 15 and 2dB km^{-1} respectively. At 183GHz, a water vapour line has an attenuation of about 35dB km^{-1}. Between these lines there are 'windows' of relatively low attenuation; e.g. around 35, 90 and 140GHz. These are the preferred bands for future exploitation of the millimetric spec-

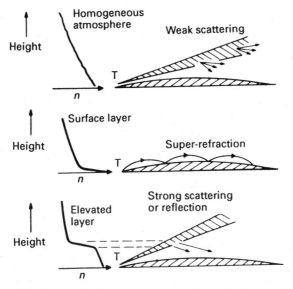

Figure 1.6   Tropospheric propagation modes and height-variation index, n

trum, such as short-range communication systems or radar. Further details are given in Report 719 of the CCIR.

1.8.1 Ground-wave terrestrial propagation

When the most distant maximum in Figure 1.5 occurs at a distance small compared with the optical range, it is often permissible to assume the earth flat and perfectly reflecting, particularly at the low-frequency end of the VHF range. The space-wave field E at a distance d is then given by Equation 1.2.

$$E = \frac{(90W^{1/2}h_t\,h_r)}{\lambda d^2} \tag{1.2}$$

W is the power radiated from a $\lambda/2$ dipole, and h_t and h_r and are the heights of the transmitting and receiving antennae respectively.The effects of irregular terrain are complex. There is some evidence that, for short, line-of-sight links, a small degree of surface roughness increases the field strength by reducing the destructive interference between the direct and ground-reflected rays. Increasing the terrain irregularity then reduces the field strength, particularly at the higher frequencies, as a result of shadowing, scattering and absorption by hills, buildings and trees. However, in the particular case of a single, obstructing ridge visible from both terminals it is sometimes possible to receive field strengths greater than those over level terrain at the same distance. This is the so-called obstacle gain.

In designing microwave radio-relay links for line-of-sight operation it is customary to so locate the terminals that, even with unfavourable conditions in the vertical gradient of refractive index (with sub-refraction effects decreasing the effective radius of the earth), the direct ray is well clear of any obstacle. However in addition to multipath fading caused by a ground-reflected ray, it is possible for line-of-sight microwave links to suffer fading caused by multi-path propagation via strong scattering or abnormal refraction in an elevated layer in the lower troposphere. This situation may lead to a significant reduction in usable bandwidth and to distortion, but the use of spaced antennae (space diversity) or different frequencies (frequency diversity) can reduce these effects. Even in the absence of well defined layers, scintillation-type fading may occasionally occur at frequencies of the order of 30-40GHz on links more than say 10km long. The development of digital systems has further emphasised the importance of studies of the effect of refractive index variation on distortion, bandwidth and error rate.As a guide to the order of magnitude of multipath fading, Report 338 of the CCIR gives the following values for a frequency of 4GHz, for the worst month of the year, for average rolling terrain in northwest Europe:

0.01% of time: path length 20km; 11dB or more
below free space
path length 40km; 23dB or more
below free space

Over very flat, moist ground the values will be greater and, for a given path length, will tend to increase with frequency. But at about 10GHz and above, the effect of precipitation will generally dominate system reliability.

The magnitude of attenuation in rain can be estimated theoretically and the reliability of microwave links can then be forecast from a knowledge of rainfall statistics. But the divergence between theory and experiment is often considerable. This is partly due to the variation which can occur in the drop-size distribution for a given rainfall rate. In addition, many difficulties remain in estimating the intensity and spatial characteristics of rainfall for a link. This is an important practical problem in relation to the possible use of route diversity to minimise the effects of absorption fading. Experimental results show that for very high reliability (i.e. for all rainfall rates less than say 50 100 mm/h in temperate climates) any terrestrial link operating at frequencies much above 30GHz must not exceed say 10km in length. It is possible, however, to design a system with an alternative route so that by switching between the two links the worst effects of localised, very heavy rain can be avoided. The magnitude of attenuation in rain is shown in Figure 1.7(a) and the principle of route diversity is illustrated in Figure 1.7(b). For temperate climates, the diversity gain (i.e. the difference between the attenuation, in dB, exceeded for a specified, small percentage of time on a single link and that exceeded simultaneously on two parallel links) varies as follows:

1.   It tends to decrease as the path length increases from 12km, for a given percentage of time and for a given lateral path separation.

2.   It is generally greater for a spacing of 8km than for 4km, though an increase to 12km does not provide further improvement.

3.   It is not strongly dependent on frequency in the range 20-40GHz for a given geometry.

The main problem in the VHF and UHF broadcasting and mobile services (apart from prediction of interference) is to estimate the effect of irregularities in terrain and of varying antenna height on the

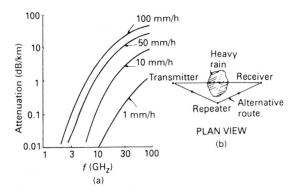

Figure 1.7 (a) Attenuation in rain; (b) the application of route diversity to minimise effects of fading

received signal. The site location is of fundamental importance. Prediction of received signal strengths, on a statistical basis, has been made using a parameter Δh, which characterises terrain roughness (see CCIR Recommendation 370 and Reports 239 and 567). However, there is considerable path-to-path variability, even for similar Δh values. Especially in urban areas, screening and multipath propagation due to buildings are important. Moreover, in such conditions, and especially for low heights of receiving antenna, depolarisation effects can impair performance of orthogonally polarised systems sharing a common frequency. At the higher UHF frequencies, attenuation due to vegetation (e.g. thick belts of trees) is beginning to be significant.

The problem of field-strength prediction becomes especially difficult at the frequencies specified for a range of future public and private mobile systems. For example, existing 'cellphone' services operating at about 950MHz will be supplemented by 'personal communication networks' (PCNs) at around 1.5 to 2.5GHz. At present, relevant data for such frequencies are very limited. Work is in progress to produce prediction models for various categories of path; e.g. over dense buildings, sparse buildings, woodland, bare ground, etc. Satellite imagery may assist in this, supplementing data on terrain height obtained from detailed maps. Preliminary results at 1.8GHz suggest that the path losses can be some 20dB or more greater than at 900MHz when trees and buildings are along the path.

1.8.2 Beyond-the-horizon propagation

Although propagation by surface or elevated layers (see Figure 1.6) cannot generally be utilised for practical communication circuits, these features remain important as factors in co-channel interference. Considerable theoretical work, using waveguide mode theory, has been carried out on duct propagation and the results are in qualitative agreement with experiment. Detailed comparisons are difficult because of the lack of knowledge of refractive index structure over the whole path, a factor common to all beyond-the-horizon experiments. Nevertheless, the theoretical predictions of the maximum wavelength trapped in a duct are in general agreement with practical experience. These values are as follows:

λ (max) in cm	Duct height in m
1	5
10	25
100	110

Normal surface ducts are such that complete trapping occurs only at centimetric wavelengths. Partial trapping may occur for the shorter metric wavelengths. Over land the effects of irregular terrain and of thermal convection (at least during the day) tend to inhibit duct formation. For a ray leaving the transmitter horizontally, the vertical gradient of refractive index must be steeper than 157 parts in 10^6 per kilometre.

Even when super-refractive conditions are absent, there remains considerable variability in the characteristics of the received signal usable in the 'scatter' mode of communication. This variability is conveniently expressed in terms of the transmission loss, which is defined as $10 \log (P_t/P_r)$, where P_t and $_r$ are the transmitted and received powers respectively. In scatter propagation, both slow and rapid variations of field strength are observed. Slow fading is the result of large-scale changes in refractive conditions in the atmosphere and the hourly median values below the long-term median are distributed approximately log-normally with a standard deviation which generally lies between 4 and 8 decibels, depending on the climate.

The largest variations of transmission loss are often seen on paths for which the receiver is located just beyond the diffraction region, while at extreme ranges the variations are less. The slow fading is not strongly dependent on the radio frequency. The rapid fading has a frequency of a few fades per minute at lower frequencies and a few hertz at UHF The superposition of a number of variable incoherent components would give a signal whose amplitude was Rayleigh-distributed. This is found to be the case when the distribution is analysed over periods of up to five minutes. If other types of signal form a significant part of that received, there is a modification of this distribution. Sudden, deep and rapid fading has been noted when a frontal disturbance passes over a link. In addition, reflections from aircraft can give pronounced rapid fading.

The long-term median transmission loss relative to the free-space value increases approximately as the first power of the frequency up to about 3GHz. Also, for most temperate climates, monthly median transmission losses tend to be higher in winter than in summer, but the difference diminishes as the distance increases. In equatorial climates, the annual and diurnal variations are generally smaller. The prediction of transmission loss, for various frequencies, path lengths, antenna heights, etc. is an important practical problem. An example of the kind of data required is given in Figure 1.8.

At frequencies above 10GHz, the heavy rain occurring for small percentages of the time causes an additional loss due to absorption,

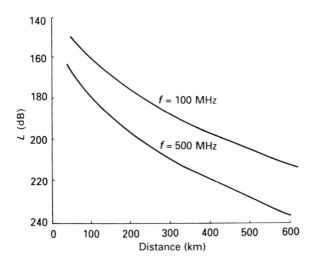

Figure 1.8   Median transmission loss, L, between isotropic antennas in a temperate climate and over an average rolling terrain. The height of the transmitting antenna is 40m, and the height of the receiving antenna is 10m

but the accompanying scatter from the rain may partly offset the effect of absorption.

1.8.3 Physical basis of tropospheric scatter propagation

Much effort has been devoted to explaining the fluctuating trans-horizon field in terms of scattering theory based on statistical models of turbulent motion. The essential physical feature of this approach is an atmosphere consisting of irregular blobs in random motion which in turn produce fluctuations of refractivity about a stationary mean value. Using this concept, some success has been achieved in explaining the approximate magnitude of the scattered field but several points of difficulty remain. There is now increasing evidence, from refractometer and radar probing of the troposphere, that some stratification of the troposphere is relatively frequent.

By postulating layers of varying thickness, horizontal area and surface roughness, and of varying lifetime it is possible in principle to interpret many of the features of tropospheric propagation. Indeed, some experimental results (e.g. the small distance-dependence of VHF fields at times of anomalous propagation) can be explained by calculating the reflection coefficient of model layers of constant height and with an idealised height-variation of refractive index such as half-period sinusoidal, exponential, etc.

The correlation between field strength and layer height has also been examined and some results can be explained qualitatively in terms of double-hop reflection from extended layers. Progress in ray-tracing techniques has also been made. Nevertheless, the problems, of calculating the field strength variations on particular links remain formidable, and for many practical purposes statistical and empirical techniques for predicting link performance remain the only solution (see CCIR Report 238).

Other problems related to fine structure are space and frequency diversity. On a VHF scatter link with antennae spaced normal to the direction of propagation, the correlation coefficient may well fall to say 0.5 for spacings of $5\text{-}30\lambda$ in conditions giving fairly rapid fading. Again, however, varying meteorological factors play a dominant role. In frequency diversity, a separation of say 3 or 4MHz may ensure useful diversity operation in many cases, but occasionally much larger separations are required.

The irregular structure of the troposphere is also a cause of gain degradation. This is the decrease in actual antenna gain below the ideal free-space value. Several aspects of the irregular refractive-index structure contribute to this effect and its magnitude depends somewhat on the time interval over which the gain measurement is made. Generally, the decrease is only significant for gains exceeding about 50dB.

1.9 Tropospheric effects in space communications

In space communication, with an earth station as one terminal, several problems arise due to refraction, absorption and scattering effects, especially at microwave frequencies. For low angles of elevation of the earth station beam, it is often necessary to evaluate the refraction produced by the troposphere, i.e. to determine the error in observed location of a satellite. The major part of the bending occurs in the first two kilometres above ground and some statistical correlation exists between the magnitude of the effect and the refractive index at the surface. For high-precision navigation systems and very narrow beams it is often necessary to evaluate the variability of refraction effects from measured values of the refractive index as a function of height. A related phenomenon important in tracking systems is the phase distortion in the wave-front due to refractive index fluctuations, a feature closely linked with gain degradation. This phase distortion also affects the stability of frequencies transmitted through the troposphere.

Absorption in clear air may affect the choice of frequencies, above about 40GHz, to minimise co-channel interference. Figure 1.9 shows the zenith attenuation from sea level for an average clear atmosphere as a function of frequency. It illustrates the 'window' regions mentioned in the survey of propagation modes. From an altitude of 4km, the values would be about one-third of those shown. This indicates the potential application of frequencies above 40GHz for communication on paths located above the lower layers of the troposphere.

Clouds produce an additional loss which depends on their liquid-water content. Layer-type cloud (stratocumulus) will not cause additional attenuation of more than about 2dB, even at 140GHz. On the other hand, cumulonimbus will generally add several decibels to the total attenuation, the exact value depending on frequency and cloud thickness.

Absorption in precipitation (see Figure 1.7(a)) has already been mentioned in relation to terrestrial systems. Water drops attenuate microwaves both by scattering and by absorption. If the wavelength

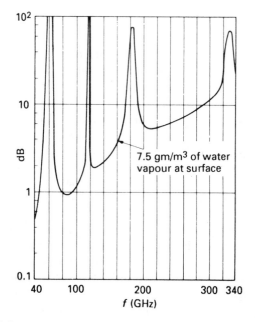

Figure 1.9 Zenith attenuation (dB) in clear air

is appreciably greater than the drop-size then the attenuation is caused almost entirely by absorption. For rigorous calculations of absorption it is necessary to specify a drop-size distribution; but this, in practice, is highly variable and consequently an appreciable scatter about the theoretical value is found in experimental measurements. Moreover, statistical information on the vertical distribution of rain is very limited. This makes prediction of the reliability of space links difficult and emphasises the value of measured data. Some results obtained using the sun as an extraterrestrial source are shown in Figure 1.10.

Scatter from rain (and ice crystals at and above the freezing level in the atmosphere) can cause significant interference on co-channel terrestrial and space systems even when the beams from the two systems are not directed towards each other on a great-circle path;

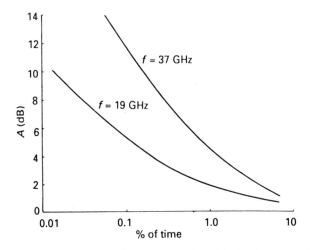

Figure 1.10 Measured probability distribution of attenuation A on earth-space path at 19GHz (southeast England: elevation angles 5° to 40°. Data from solar tracking radiometers)

such scattering being, to a first approximation, isotropic. It may also be significant in the case of two earth stations with beams elevated well above the horizon: for example, with one a feeder link transmitter to a broadcasting satellite and the other a receiver in the fixed-satellite service. This mode of interference may be dominant when hills or other obstacles provide some 'site-shielding' against signals arriving via a ducting mode.

Because precipitation (and to a smaller extent atmospheric gases) absorb microwaves, they also radiate thermal-type noise. It is often convenient to specify this in terms of an equivalent blackbody temperature or simply noise temperature for an antenna pointing in a given direction. With radiometers and low-noise receivers it is now possible to measure this tropospheric noise and assess its importance as a factor in limiting the performance of a microwave earth space link. For a complete solution, it is necessary to consider not only direct radiation into the main beam but also ground-reflected radiation, and emission from the ground itself, arriving at the receiver via side and back lobes. From the meteorological point of view, radiometer soundings (from the ground, aircraft, balloons or satellites) can provide useful information on tropospheric and stratospheric structure.

Absorption in precipitation becomes severe at frequencies above about 30GHz and scintillation effects also increase in importance in the millimetre range. However, for space links in or near the vertical direction the system reliability may be sufficient for practical application even at wavelengths as low as 3-4mm. Moreover, spaced receivers in a site-diversity system can be used to minimise the effects of heavy rain.

In recent years, extensive studies of propagation effects (attenuation, scintillation, etc.) have been carried out by direct measurements using satellite transmissions. Special emphasis has been given to frequencies between 10 and 30GHz, in view of the effect of precipitation on attenuation and system noise. Details are given in Report 564 of the CCIR. For typical elevation angles of 30° to 45°, the total attenuation exceeded for 0.01% of the time has values of the following order:

f=12GHz:	5dB in temperate climates
	(e.g. northwest Europe)
	20dB in tropical climates (e.g. Malaysia)
	10dB in East Coast USA (Maryland)
f=20GHz:	10dB (northwest Europe)

Site-diversity experiments using satellite transmissions show that site spacings of the order of 5-10km can give a useful improvement in reliability. However, the improvement may depend on the site geometry and on topographical effects. At frequencies above say 15GHz, the advantage of site diversity may be quite small if the sites are so chosen that heavy rain in, for example, frontal systems tends to affect both sites simultaneously.

Frequency re-use is envisaged, in space telecommunication systems, by means of orthogonal polarisation. But this technique is restricted by depolarisation due to rain and ice clouds and, to a lesser extent, by the system antennas. Experimental data on polarisation distortion, obtained in satellite experiments, are given in CCIR Report 564.

Further data are still needed, especially from satellite transmissions and at frequencies up to at least 40GHz. In this context, the availability of signals from the Olympus satellite at 12, 20 and 30GHz, from 1989 onwards, is important. Measurements at 20, 40 and 50GHz on the ITALSAT satellite are also planned. But an ongoing need still remains for improved predictions for reception in low-latitude (equatorial) regions. This has recently been the subject of study in a Working Party of the CCIR.

1.10 Propagation and co-channel interference

The increasing need for different microwave terrestrial systems, satellite communications and broadcast systems to share frequencies has produced a correspondingly greater likelihood of mutual interference. The development of interference prediction methods leading to the efficient co-ordination of communication systems sharing a frequency band requires modelling the transmission loss due to all possible propagation mechanisms as a function of frequency, distance and time percentage. Until recently, the primary concern had been for situations occurring less than 1% of time, for which the main clear-air mechanisms are ducting and layer reflection. However the small antenna systems used in new services have generated a need for time percentages up to 20%, for which diffraction and tropospheric scatter are important.

In addition, scattering from different types of hydrometers can be important at frequencies above about 10GHz. This propagation mechanism may couple energy into the receiving antennas of unrelated systems, despite precautions such as site-shielding. Up to now there have been only very limited data available to develop prediction models.

A third group of studies in response to this general area has reflected that an accurate estimation of diffraction effects due to buildings and terrain is essential in the planning of radio services in the microwave bands. Significant data exist from diffraction over large-scale features of terrain, e.g. hills and mountains, which are of major concern at VHF and UHF, but little data are available on the diffraction losses for microwaves that can be produced by buildings or trees which could be used to shield antennas from interference signals.

All these aspects have been studied in recent years in a collaborative European Project 'COST 210' (Influence of the Atmosphere on Interference Between Radiocommunication Systems at Frequencies Above 1GHz). The full report should be consulted for details of the procedures developed.

1.11 Techniques for studying tropospheric structures

The importance of a knowledge of the structure of the troposphere in studies of propagation is clearly evident in the above sections. The small-scale variations in refractive index and in the intensity of precipitation are two important examples. They form part of the general topic of tropospheric probing.

Much useful information on the height-variation of refractive index can be obtained from the radio-sondes carried on free balloons and used in world-wide studies of meteorological structure. However, for many radio applications these devices do not provide sufficient detail. To obtain this detail instruments called refractometers have been developed, mainly for use in aircraft, on captive balloons or on tall masts. They generally make use of a microwave cavity for measuring changes in a resonance frequency, which in turn is related to the refractive index of the enclosed air. Such refractometers are robust, rapid-response instruments which have been widely used as research tools, though they have yet to be developed in a form suitable for widespread, routine use.

High-power, centimetric radar is also a valuable technique. By its use it is possible to detect layers or other regions of strong scatter in the troposphere, and to study their location and structure. Joint radar-refractometer soundings have proved of special interest in confirming that the radar does indeed detect irregularities in clear-air structure. The application of radar in precipitation studies is, of course, a well-known and widely used technique in meteorology; although to obtain the detail and precision necessary for radio applications requires careful refinements in technique.

Optical radar (lidar) and acoustic radar have also been used to probe the troposphere, although the information they provide is only indirectly related to radio refractive index.

The millimetre and sub-millimetre spectrum, as yet not exploited to any significant degree for communications, is nevertheless a fruitful region for tropospheric probing. In particular, the presence of several absorption lines (in water vapour, oxygen and minor constituents such as ozone) makes it possible to study the concentration and spatial distribution of these media. Near the ground, direct transmission experiments are feasible; for example, to study the average water-vapour concentration along a particular path. In addition, it is possible to design radiometers for use on the ground, in an aircraft or in a satellite, which will provide data on the spatial distribution of absorbing atmospheric constituents by measurement of the emission noise they radiate. This topic of remote probing is one exciting considerable current interest in both radio and meteorology.

1.12 References

Aarons, J. (1973) *Total-electron content and scintillation studies of the ionosphere*, ed. AGARDograph 166, NATO, Neuilly- sur-Seine, France.

Anderson, S.J. (1986) Remote sensing with the JINDALEE skywave radar, *IEEE J. Ocean. Eng.*, OE-11(2), 158.

Appleton, E.V. and Beynon, W.J.G. (1940), (1947) The application of ionospheric data to radio communications, *Proc. Phys. Soc.* **52**. 518 and **59**. 58.

Appleton, E.V. and Piggott, W.R. (1954) Ionospheric absorption measurements during a sunspot cycle, *J. Atmosph. Terr. Phys.* **5**. 141.

Bailey, D.K. (1964) Polar cap absorption, *Planet. Space Sci.* **12**. 495.

Bailey, D.K., Bateman, R. and Kirby, R.C. (1955) Radio transmission at VHF by scattering and other processes in the lower ionosphere, *Proc. IRE,* **43**. 1181.

Baker, D.M. and Davies, K. (1968) Waves in the ionosphere produced by nuclear explosions, *J. Geophys. Res.* **73**. 448.

Baker, D.M. and Davies, K. (1969) F2-region acoustic waves from severe weather, *J. Atmosph. Terr. Phys.* **31**. 1345.

Bean, B.R. and Dutton, E.J. (1966) *Radio meteorology*, Monograph 92, US Government Printing Office, Washington.

Bello, P.A. (1965) Some techniques for instantaneous real-time measurements of multipath and Doppler spread. *IEEE Trans. Comm. Tech.* **13**. 285.

Belrose, J.S. and Burke, M.J. (1964) Study of the lower ionosphere using partial reflections, *J. Geophys, Res.* **69**. 2799.

Beynon, W.J.G. (1967) *Special issue on analysis of ionograms for electron density profiles*, ed. URSI Working Group, *Radio Science* **2**. 1119.

Beynon, W.J.G. and Rangaswamy, S. (1969) Model electron density profiles for the lower ionosphere, *J. Atmosph. Terr. Phys.* **31**. 891.

Blair, J.C., Melanson, L.L. and Tveten, L.H. (1969) HF ionospheric radar ground scatter map showing land-sea boundaries by a spectral separation technique, *Electronics Letters*, **5**. 75.

Bradley, P.A. and Dudeney, J.R. (1973) A simple model of the vertical distribution of electron concentration in the ionosphere, *J. Atmosph. Terr. Phys.* **35**. 2131.

Boithias, L. (1988) *Radio-wave propagation*, North Oxford Academic Publishers, London.

Buchau, J. (1972) *Instantaneous versus averaged ionosphere*. Air Force Surveys in Geophysics No. 241 (Air Force Systems Command, United States Air Force), 1.

Budden, K. (1961) *Radio waves in the ionosphere*, Cambridge University Press, Cambridge.

Castel, F. Du, (1966) *Tropospheric radiowave propagation beyond the horizon*, Pergamon Press, Oxford.

CCIR (1990) REPORT 266-7. *Ionospheric propagation and noise characteristics pertinent to terrestrial radiocommunication systems design and service planning (Fading)*, Documents of XVIIth Plenary Assembly, Dusseldorf, ITU, Geneva.

Cohen, R. and Bowles, K.L. (1961) On the nature of equatorial spread-F. *J. Geophys. Res.* **66**. 1081.

Coll, D.C. and Storey, J.R. (1964) Ionospheric sounding using coded pulse signals, *Radio Science.* **68D**. 1155.

Croft, T.A. (1968) Special issue on ray tracing, *Radio Science.* **3**. 1.

Croft, T.A. (1972) Skywave backscatter: a means for observing our environment at great distances, *Rev. Geophys. and Space Physics.* **10**. 73.

Croft, T.A. and Hoogasian, H. (1968) Exact ray calculations in a quasi-parabolic ionosphere with no magnetic field, *Radio Science.* **3**. 69.

Davies, K. (1969) *Ionospheric radio waves*. Blaisdell, Waltham, Mass.

Davies, K. (1990) *Ionospheric radio*. Peter Peregrinus, London.

Davies, K. and Baker, D.M. (1965) Ionospheric effects observed around the time of the Alaskan earthquake of March 28 1964. *J. Geophys. Res.* **70**. 2251.

Duncan, R.A. (1960) The equatorial F-region of the ionosphere, *J. Atmosph. Terr. Phys.* **18**. 89.

Dyson, P.L. and Bennett, J.A. (1988) A model of the vertical distribution of the electron concentration in the ionosphere and its application to oblique propagation studies. *J. Atmosph. Terr. Phys.* **50**(3). 251.

EEC (1990) COST Project 210; *Influence of the atmosphere on interference between radio systems at frequencies above 1GHz*. (L- 2920, Luxembourg).

Fejer, J.A. (1955) The interaction of pulsed radio waves in the ionosphere, *J. Atmosph. Terr. Phys.* **7**. 322.

Fenwick, R.B. and Barry, G.H. (1966) Sweep frequency oblique ionospheric sounding at medium frequencies, *IEEE Trans. Broadcasting*, **12**. 25.

Fenwick, R.B. and Villard, O.G. (1963) A test of the importance of ionosphere reflections in long distance and around-the-world high frequency propagation, *J. Geophys. Res.* **68**. 5659.

Garriott, G.K. and Nichol, A.W. (1965) Ionospheric information deduced from the Doppler shifts of harmonic frequencies from earth satellites, *J. Atmosph. Terr. Phys.* **22**. 50.

Hall, M.P.M. (1979) *Effects of the troposphere on radio communication*, Peter Peregrinus (for IEE) London.

Hall, (1989) *Radiowave propagation*; edited by M.P.M. Hall and L.W. Barclay. IEE Electromagnetic Waves Series 30, London.

Hargreaves, J.K. (1969) Auroral absorption of H.F. radio waves in the ionosphere — a review of results from the first decade of riometry. *Proc. IEEE*, **57**. 1348.

Hartz, T.R. (1968) The general pattern of auroral particle precipitation and its implications for high latitude communication systems, in *Ionospheric Radio Communications,* ed. K. Folkestad, Plenum, New York, 9.

Haselgrove, J. (1954) Ray theory and a new method for ray tracing. In *Report on conference on physics of ionosphere*, Phys. Soc. London, 355.

Headrick, J.M. and Skolnik, M.I. (1974) Over-the-horizon radar in the HF band. *Proc. IEEE* **62**(6). 664.

Helliwell, R.A. (1965) *Whistlers and Related Ionospheric Phenomena*, Stanford University Press, Stanford, California.

ITU-R (1995a) Recommendation PI 531, *Ionospheric effects influencing radio systems involving spacecraft*, ITU, Geneva.

ITU-R (1995b) Recommendation PI 532, *Ionospheric effects and operational considerations associated with artificial modifications of the ionosphere and the radio-wave channel*, ITU, Geneva.

ITU-R (1995c) Recommendation PI 684, *Prediction of field strength at frequencies below about 500kHz*, ITU, Geneva.

ITU-R (1995d) Recommendation PI 434, *ITU-R reference ionospheric characteristics and methods of basic MUF, operational MUF and ray-path prediction*, ITU, Geneva.

ITU-R (1995e) Recommendation PI 533, *HF propagation prediction method*, ITU, Geneva.

Matsushita, S. (1967) Geomagnetic disturbances and storms, in *Physics of geomagnetic phenomena*, ed. Matsushita, S. and Campbell, W.H., Academic Press, London, 793.

Mitra, S.N. (1949) A radio method of measuring winds in the ionosphere, *Proc. IEE*, **46** Pt.III, 441.

Muldrew, D.B. (1965) F-layer ionisation troughs deduced from Alouette data. *J. Geophys. Res.* **70**.2635.

Munro, G.H. (1950) Travelling disturbances in the ionosphere, *Proc. Roy. Soc.* **202A**. 208.

Nielson, D.L. (1968) The importance of horizontal F-region drifts to transequatorial VHF propagation, *Scatter propagation of radio waves*, ed Thrane, E. *AGARD Conference Proceedings*, No.37, NATO, Neuilly-sur-Seine, France.

Pierce, E.T. (1965) Nuclear explosion phenomena and their bearing on radio detection of the explosions, *Proc. IEEE*, **53**. 1944.

Pierce, J.A. (1965) OMEGA, *IEEE Trans. Aer. and Elect. Syst.* 1. 206.

Piggott, W.R. and Rawer, K. (1972) *U.R.S.I. handbook of ionogram interpretation and reduction*, 2nd edn. Rep. UAG-23, Dept. of Commerce, Boulder, USA.

Ratcliffe, J.A. (1956) *Some aspects of diffraction theory and their application to the ionosphere*, Reports on Progress in Physics, Phys. Soc., London, **19**. 188.

Ratcliffe, J.A. (1959) *The magnetoionic theory*, Cambridge University Press.

Ratcliffe, J.A. (1970) *Sun, Earth and Radio — an introduction to the ionosphere and magnetosphere*, Weidenfeld and Nicolson, London.

Ratcliffe, J.A. (1972) *An introduction to the ionosphere and magnetosphere*, Cambridge University Press, Cambridge.

Reinisch, B.W. and Huang Xueqin (1983) Automatic Calculation of electron density profiles from digital ionograms, 3. Processing of bottomside ionograms. *Radio Science*, **18**(3). 477.

Rishbeth, H. (1972) Thermospheric winds and the F-region, a review, *J. Atmosph. Terr. Phys.* **34** 1.

Rishbeth, H and Garriott, O.K. (1969) *Introduction to ionospheric physics*, Academic Press, London.

Ross, W.J. (1965) Second-order effects in high frequency transionospheric propagation, *J. Geophys. Res.* **70**. 597.

Rutherford (1991) *Catalogue of ionospheric vertical sounding data*, World Data Centre C1 for Solar-Terrestrial Physics, Rutherford Appleton Laboratory, Chilton, England.

Saxton, J.A.(Ed) (1964) *Advances in radio research*, Academic Press, London.

Smith, E.K. (1970) *Electromagnetic probing of the upper atmosphere*, U.R.S.I. Working group, *J. Atmosph. Terr. Phys.* **32**. 457.

Smith, E.K. and Matsushita, S. (1962) *Ionospheric sporadic- E.* Macmillan, New York.

Stevens, E.E. (1968) *The CHEC sounding system, in Ionospheric radio communications*, ed. K. Kolkestad, Plenum, New York, 359.

Stringer, F.S. (1969) Hyperbolic radionavigation systems, *Wireless World* **75**. 353.

Sugar, G.R. (1964) Radio propagation by reflection from meteor trails, *Proc. IEEE* **52**. 116.

Teters, L.R., Lloyd, J.L., Haydon, G.W. and Lucas, D.L. (1983) *Estimating the performance of telecommunication systems using the ionospheric transmission channel — Ionospheric Communications Analysis and Prediction Program,* Report 83-127, National Telecommunications and Information Administration, Dept. of Commerce, Boulder, U.S.A.

Titheridge, J.E. (1985) *Ionogram analysis with the generalised program POLAN*, Rep. UAG-93, Dept. of Commerce, Boulder, U.S.A.

URSI Commission F. Colloquium Proceedings; La Baule, France (CNET, Paris, 1977). Also at Lennoxville, Canada (Proceedings edited by University of Bradford, England, 1980).

Whalen, J.A., Buchau, J. and Wagner, R.A. (1971) Airborne ionospheric and optical measurements of noontime aurora, *J. Atmosph. Terr. Phys.* **33**. 661.

2. Radiowave propagation

2.1 Introduction

Electromagnetic energy radiates outwards from the source, usually an antenna, at approximately the speed of light and it is attenuated and influenced by the medium through which it travels. Radio communications necessitates launching RF energy into the propagation medium, detecting its presence at some remote point and recovering the information contained within it, while eliminating noise and other adverse factors introduced over the transmission path. An understanding of radio wave propagation is therefore essential in the planning and operation of radio communications systems, to ensure that communications can be established and that there is an optimum solution between costs (capital and running costs) and link availability.

This chapter examines radio wave propagation in the frequency bands from VLF (10kHz) to the millimetric band (100GHz), and the influence of the earth, the atmosphere and the ionosphere on such transmissions. It is assumed that propagation is in the far field i.e. several wavelengths from the antenna, where the electric and magnetic components of the wavefront are at right angles and normal to the direction of propagation. The process by which energy is launched into the propagation medium in the near field is described in Chapter 4, on antennas.

Radio waves may be propagated in one or more of five modes, depending upon the medium into which they are launched and through which they pass. These modes are:

1. Free space propagation, where radio waves are not influenced by the earth or its atmosphere.
2. Ground wave propagation, where radio waves follow the surface of the earth.

3. Ionospheric propagation, where radio waves are refracted by ionised layers in the atmosphere.
4. Tropospheric propagation, where transmission is 'line of sight' with some atmospheric refraction occurring.
5. Scatter propagation, where natural phenomena such as tropospheric turbulence or ionised meteor trails are used to scatter radio waves.

2.2 The radio frequency spectrum

The radio frequency spectrum is divided into a number of bands which have been given designations such as LF, MF, HF etc. for ease of reference. These are shown in Figure 2.1 and are presented on a logarithmic scale. The microwave band, usually taken to be from

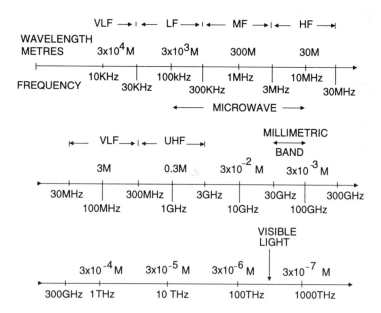

Figure 2.1 Radio frequency bands and designations

1GHz to 30GHz has been subdivided into a number of sub-bands which have been given letter designations such as X band (8GHz to 12GHz), but differing definitions are used and can cause confusion. 30GHz to 100GHz is commonly referred to as the millimetric band. Allocations have been made up to 275GHz by the ITU, but there is little activity other than experimental work above 100GHz.

Frequency and wavelength are shown in Figure 2.1. As radiowaves propagate at 3×10^8 metres per second in free space, frequency and wavelength are related by Equation 2.1, where λ is wavelength in metres, and f is frequency in cycles per second.

$$\lambda \times f = 3 \times 10^8 \tag{2.1}$$

2.3 Free space propagation

This is seldom a practical situation, but it occurs when both transmitting and receiving antennas are situated away from the influence of the earth's surface or other reflecting and absorbing objects, including the transmitter and receiver themselves. If power is fed to an isotropic antenna, i.e. an antenna that radiates equally in all directions in azimuth and elevation, the wave front will radiate outwards from the antenna in an ever expanding sphere at 3×10^8 metres per second (186,282 miles per second). The strength of the signal obviously decreased with distance as a given amount of power is spread over a greater area, and the incident power density at a remote point can be calculated as in Equation 2.2 (CCIR, 1978), where P_r is received power density in watts per meter^2, P_t is transmitted power in watts and r is distance in metres.

$$P_r = \frac{P_t}{4 \pi r^2} \tag{2.2}$$

Because of the very large differences in power density over long propagation paths, particularly in the microwave bands, it is usual to measure power density in decibels (dB) relative to 1 watt i.e. dBW, or 1 milliwatt i.e. dBm.

At frequencies below the microwave range, field strength in volts per metre is a more usual measurement than power density, partly because volts per metre has historically been used for the measurement of signal strength and partly because the thermal heating effect of power absorption is easier to use for microwave measurements. The conversion can be made for power density to field strength using the simple derivation of Ohm's law (Braun, 1986) as in Equation 2.3, where V is voltage, R is resistance in ohms and P is power in watts.

$$V^2 = R P \qquad (2.3)$$

In this case, R is substituted for Z_o, the characteristic impedance of free space, which is a constant of 377 ohms.

Field strength can therefore be converted to power density by (Maslin, 1987) Equation 2.4, where E is field strength in volts per metre, Z_o is the impedance of free space (377 ohms), P_d is power density in watts per metre, and P_d is power density in watts per metre^2.

$$E^2 = Z_o P_d \qquad (2.4)$$

Equation 2.5 can be used to convert directly from dBW to dBμV, where E is in dB.

$$E = P_d \; dB \; (1W/m^2) + 145.6 \qquad (2.5)$$

On the left hand side of this equation, the dB represents a voltage and not a power ratio.

The voltage at the centre of an un-terminated half wave dipole in the path of a radiowave and aligned along the axis of the electric vector (i.e. no polarisation loss) is given by Equation 2.6, where V is the voltage at the centre of the dipole, E is the field strength in volts per metre, and λ is the wavelength in metres.

$$V = \frac{E\lambda}{\pi} \tag{2.6}$$

If the dipole is connected to a feeder of matching impedance, E is halved. If polarisation and feeder losses are taken into account, the input to a receiver can be calculated.

Free space attenuation, derived from the power density Equation 2.1 and from 2.6, and published in CCIR Report 252-2 is given by Equation 2.7, where L_{fr} is the free space loss in decibels, f is the frequency in megahertz and d is distance in kilometres.

$$L_{fr} = 32.44 + 20\log f + 20\log d \tag{2.7}$$

It will be noted that Equation 2.7 introduces a frequency component into the calculation.

As an example, the power density from a 500 watt transmitter working into a lossless isotropic antenna at a distance of 1 kilometre is $0.039W^{-3}/m^2$, and at 16 kilometres is $0.155W^{-6}/m^2$. This is a power ratio of 24dB, which is to be expected as there is a 6dB power loss every time the distance is doubled. The CCIR formula (Equation 2.7) gives a free space loss of 72.4dB at 1km and 96.5dB at 16km, again a difference of 24dB. Figure 2.2 shows the free space loss for frequency and distance based on this formula.

2.4 The propagation medium

Almost all propagation involves transmission through some of the earth's atmosphere, and the structure of the atmosphere and features within it are shown in Figure 2.3. Most meteorological activity and cloud formation occur in the first 10km, and jet aircraft cruise at between 10km and 15km. Air pressure falls with height and at 30km, radiation from the sun is sufficient to generate some free electrons, but the first distinct ionised layer, the D layer occurs at 70km. Above the D layer, temperature and incident radiation increase and the E, F_1 and F_2 layers are formed between 120km and 450km. The ionosphere, which is the region in which the ionised layers are formed,

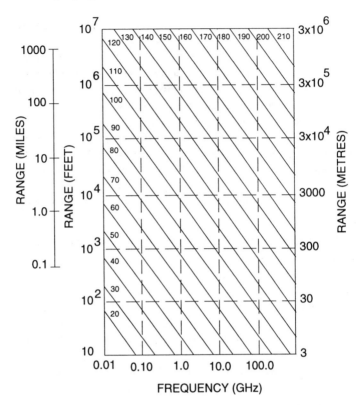

Figure 2.2   Free space loss versus distance for frequencies from 1MHz to 100GHz. (The units on the diagonal lines are in dB)

spans the region from 50km to 600km. Ionised trails from meteors occur at around 100km and the lowest satellite orbit is at about 150km.

The primary influence on the atmosphere is radiation from the sun, which causes the ionised layers to form and creates global climatic and regional weather patterns. Solar radiation follows an 11 year sun spot cycle, which has a direct correlation with ionisation and is recorded on a scale from 0 to about 200. This is a smoothed average of the number of sunspots, which are observed disturbances on the sun's surface. A sunspot number is issued by the Sunspot Data Centre

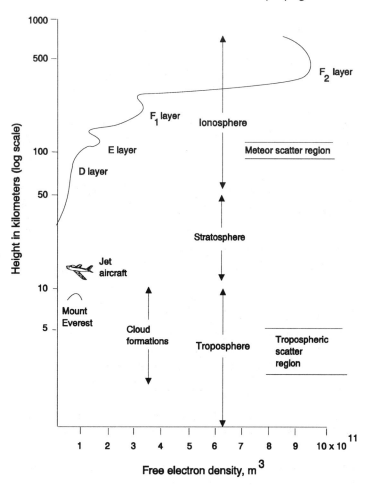

Figure 2.3 The Earth's atmosphere; major features and ionisation

in Brussels and also the Telecommunications Services Centre in Boulder, Colorado, U.S.A. and is a smoothed 12 month average of solar activity indicating the degree of ionisation that can be expected. High sunspot numbers are an indication of better conditions for long range HF communication and this may also lead to unwanted long distance propagation in the VHF bands creating interference in mobile communications and FM and TV broadcasting services.

The sunspot number being a running average may not be an accurate indication of daily or hourly conditions. The solar flux is a measure of solar activity taken at 2.8GHz and is a better indication of real time ionospheric conditions. It is measured continuously and quoted on a scale usually in the range from 60 to 260. The US National Bureau of Standards radio station WWV transmits ionospheric data information hourly.

The signal to noise ratio (S/N) is often the determining factor in establishing communications, and it depends upon the absolute level of signal, the external noise in the propagation medium and the internal noise in the transmitting and receiving equipment. At HF, communications are normally externally noise limited so that it is the noise in the propagation medium which predominates, whilst at VHF and above the noise generated within the first stages of the receiver is usually the determining factor. Therefore there may be no benefit in trying to increase the received strength of HF signals by using larger and higher gain antennas as noise may increase proportionally with no improvement in signal to noise ratio.

External noise is from three sources; galactic noise, atmospheric noise and man made noise. Noise power in a communication link is given by Equation 2.8, where N is the noise power in watts, T is the temperature in degrees absolute, often taken as 290° as this corresponds to the normal room temperature of $17^{o}C$ and B is Boltzmanns constant (1.38×10^{-23}).

$$N = k\,T\,B \tag{2.8}$$

In communications links, noise is measured in terms of the noise power available from a lossless antenna and may be expressed as in Equation 2.9, where P_n is the total power in dBW, F_a is the antenna noise figure in dB, B = 10log bandwidth (in hertz), and 204 is 10log kT^{o}, assuming T^{o} = 290K.

$$P_n = F_a + B - 204 \tag{2.9}$$

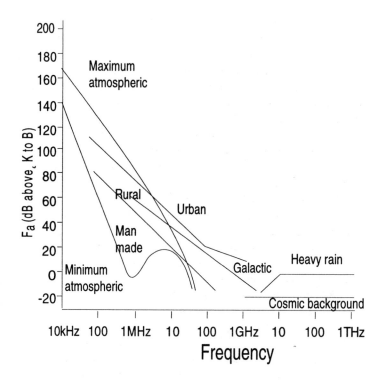

Figure 2.4   Noise in the propagation medium; typical figures

Atmospheric and man made noise information is given in CCIR publications and typical values are shown in Figure 2.4. Galactic noise originates from sources outside the Earth's atmosphere, such as the sun and the stars, and extends from about 15MHz to 100MHz. It is limited by ionospheric absorption below this frequency range and atmospheric absorption above it. Atmospheric noise is the major source of noise in the MF and HF bands and is primarily due to lightning discharges so it is particularly severe in the rainy season in tropical regions such as equatorial Africa, and at its lowest value in high latitudes at night. It is transmitted over long distances by sky-wave paths. Man made noise may be similarly transmitted and emanates from power lines, industrial machinery and fluorescent tubes. Four standard levels for business, residential, rural and quite rural sites are defined in the CCIR report (CCIR, 1978).

Analysis of a propagation path usually requires a calculation of distance from transmitter to receiver and it is useful to know the great circle bearing. This information can be calculated from Equations 2.10 and 2.11 for two points A and B on the surface of the Earth, where A is point A latitude in degrees; B is point B latitude in degrees; L is the difference in longitude between A and B; C is the true bearing of the receiver from the transmitter, which may be $360^{\circ} - C$ if a negative value is calculated; D is the distance along the path in degrees of arc, which may be converted to kilometres by multiplying by 111.111 (i.e. 1 degree of arc = 111.111 kilometres).

$$\cos D = \sin A \sin B + \cos A \cos B \cos L \qquad (2.10)$$

$$\cos C = \frac{\sin B - \sin A \cos D}{\cos A \sin D} \qquad (2.11)$$

2.5 Low and medium frequency ground waves

In the low and medium frequency bands up to 3MHz, ground wave (rather than sky wave) propagation is used, because sky wave is heavily absorbed in daytime by the D layer. LF and MF antennas are generally short in terms of wavelength so vertical mast radiators are used. These have the advantages of maximum radiation at low angles to the horizon and also radiating vertically polarised transmissions.

The attenuation of vertically polarised ground wave transmissions is very much less than for horizontal polarisation, for example the ground wave attenuation of a 2MHz vertically polarised transmission over medium soil is 45dB at 30km whereas a horizontally polarised signal would be attenuated by nearly 95dB. Vertical polarisation is therefore almost always employed.

Because currents are induced in the ground by vertically polarised ground wave transmissions, attenuation is also dependent on ground conductivity and dielectric constant. Salt water offers low attenuation whereas desert sand or polar ice offer high attenuation, and consequent reduced communications coverage. Terrain irregularities such as hills and mountains, and also rough (as compared with calm) seas also reduce coverage.

Received field strength at any distance can be calculated theoretically (Braun, 1986; Maslin, 1987), however CCIR publish a series of graphs showing received field strength in microvolts per metre and in dBs relative to 1 microvolt, for distances up to 1000km. Frequencies from 500khz to 10MHz, and homogeneous surface conditions from sea water to dry soil are shown. Some of the curves are shown in Figure 2.5 and are calculated on the basis of a 1kW transmitter

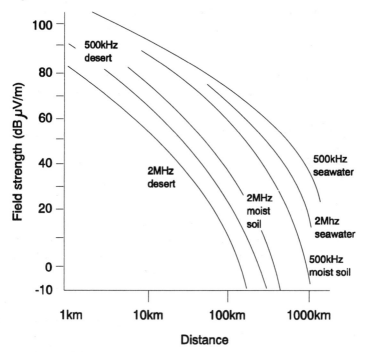

Figure 2.5 Ground wave propagation; field strength for different frequencies and surface conditions

working into a short omni-directional monopole antenna. The field strength values are proportional to the square root of the power and will have to be adjusted accordingly for different powers or for higher gain or directional antennas.

The CCIR curves shown in Figure 2.5 assume propagation over homogeneous terrain, but discontinuities occur in paths over land and sea or over different types of soil, and the field strength over such a path can be calculated by a method developed by Millington, 1949. An example of the field strength over a land and sea path is shown in Figure 2.6 and it is interesting to note that the field strength rises after the wave front passes from land to sea.

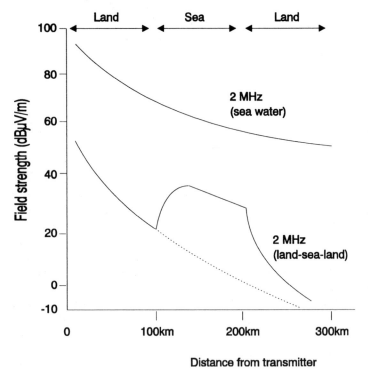

Figure 2.6 Ground wave field strength curve for a hypothetical non-homogeneous path

MF groundwave propagation has the advantage of predictable but limited communications coverage, which is largely independent of ionospheric conditions and diurnal or seasonal variations. Up to 1000km are achievable over sea water, but in desert conditions this may be limited to tens of kilometres unless very high powers and directional antennas are used. A limitation that often occurs in medium frequency broadcasting is interference between groundwave and skywave from the same transmitter during darkness due to the absence of the absorbing D layer causing deep and rapid fading in received signals. The only practical solution is limiting power radiated at night when the D layer is no longer present, and designing the antenna to minimise high angle radiation. At LF and VLF, very long distance and global communications can be achieved, but this necessitates the use of high transmitter powers and very large antenna systems.

2.6 HF skywave propagation

Propagation in the HF band from 3MHZ to 30MHz is probably the most variable and least predictable of all transmission modes, as it depends upon the height and intensity of the ionised layers in the ionosphere. A wealth of data has been collected and incorporated in comprehensive computer programmes to enable predictions to be made with a reasonable degree of statistical accuracy, and modern ionospheric sounders can evaluate conditions on a real time basis. Unpredictable events such as solar flares do however mean that HF propagation always contains some degree of uncertainty (as does most radio wave propagation), and predictions can only be made on a statistical probability basis.

The ionosphere which is the primary influence on HF propagation extends from approximately 80km to 300km above the Earth's surface, and divides into a number of distinct ionised layers. Variations in the height and intensity of the layers occurs on a diurnal and seasonal basis due to the rotation and position of the earth in relation to the sun, and also on the longer term 11 year sunspot cycle. The diurnal variation is shown in Figure 2.7, and it will be seen that at night time when incident radiation is at a minimum as the ionosphere

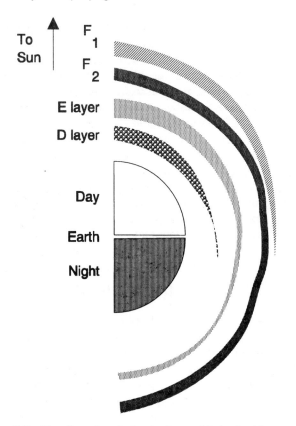

Figure 2.7   The diurnal variation in the earth's ionised layers

is in the earth's shadow, the ionosphere comprises two comparatively thin layers at 110km and 210km, called the E and F layer respectively. During daylight, these two layers increase in thickness and intensity, and the F layer divides into the two separate F_1 and F_2 layers at 210km and 300km. In addition, the D layer forms at 80km during daylight but disappears at night. Similar variations occur with the seasons so there is a greater level of ionisation in northern latitudes during July (summer) than in January (winter), but the reverse applies in the southern hemisphere.

The D layer actually spans an altitude from 50km to 90km and absorbs frequencies in the MF and lower HF bands. Higher frequencies will pass through the D layer suffering some attenuation and may be reflected back to earth by the E layer at distances up to 2000km. Still higher frequencies within the HF band will pass through the E layer and may be reflected by one of the F layers, providing very long distance communications. Frequencies in the VHF bands and above (i.e. above 30MHz) will generally pass through all the layers and into space, except in unusual conditions of strong ionospheric activity.

Although a radio wave can be visualised as being reflected by the ionosphere, the process is in fact one of refraction, and the angle of refraction is proportional to both the angle of incidence and the frequency. When the wave front enters an ionised layer, it excites free electrons into oscillation which re-radiate electromagnetic energy, and this re-radiation modifies the direction of the wavefront, tending to bend it back to earth. For a given frequency, this tendency increases as the angle of incidence is reduced, so at the critical angle, the wave will be refracted back to earth, but at greater angles (i.e. nearer the vertical) only partial or no refraction will occur. This is illustrated in Figure 2.8.

Similarly, for a given angle of incidence, refraction will decrease with increase in frequency, so at a critical frequency the wavefront will pass through the layer. As the wavefront passes through an ionised layer, it imparts energy to the electrons and a small amount of this energy is lost as heat. In the D layer, this transfer of energy is sufficient to completely absorb medium frequencies during daylight but at higher frequencies it causes some attenuation in the refraction process. Refracted waves also undergo a change of polarisation due to the complex movement of the free electrons, and this phenomenon is called Faraday rotation.

In planning an HF skywave link, it will be necessary to know the distance between the transmitter and receiver, or the area to be covered, and this can be calculated by Equations 2.10 and 2.11. It is desirable to minimise the number of 'hops' i.e. refractions from the ionosphere and reflections from the earth's surface, to minimise path attenuation and the inherent variability associated with ionospheric refraction. Maximum radiation from the antenna should therefore be

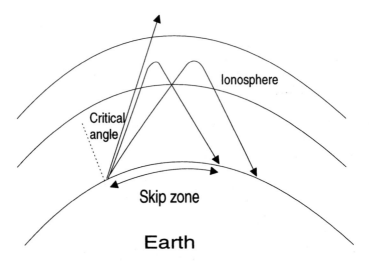

Figure 2.8 The critical angle of ionospheric refraction

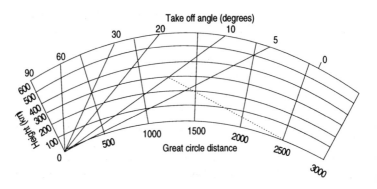

Figure 2.9 Take off angle for single hop sky wave paths

at an angle (the take off angle) that results in refraction from the ionosphere onto the target area. Figure 2.9 is a diagrammatical representation of the surface of the earth from which the optimum take off angle of radiation from the antenna can be derived if the distance and

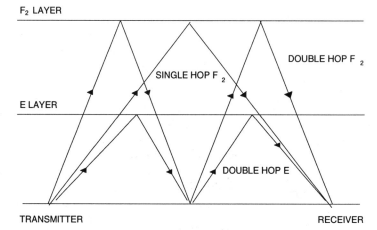

Figure 2.10 Possible multi-hop sky wave paths

refracting layer height are known. A lower take off angle is required for longer distances and for distance of greater than 2000km, two or more hops will normally be required.

In practice, a number of different propagation modes may be established on multi-hop links, which makes field strength predictions for the receive site difficult and uncertain. Three possible modes are shown in Figure 2.10. A first approximation can be obtained by calculating the free space loss using Equation 2.7, taking into account increased distance due to ionospheric refraction and applying this to the field strength at the transmitter site. Additional losses of up to 20dB will result from ionospheric and ground reflections, but these are very approximate values.

Actual sky wave refraction losses can be calculated (Braun, 1986), as can received field strength (Damboldt, 1975). Computer propagation prediction programmes are now generally employed for analysing HF paths and predicting received signal strength because of the large number of cancellations required, rather than their inherent complexity. Such programmes run on desktop PC computers (Hitney, 1990) although many research establishments have larger and more sophisticated programmers (Dick, 1990).

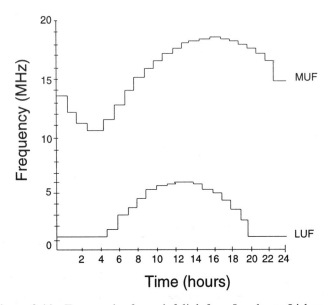

Figure 2.11 Frequencies for an h.f. link from London to Lisbon

Figure 2.11 shows a typical printout of a link from London to Lisbon, a great circle distance of 1656 kilometres and a single hop F2 distance of 1757 kilometres. The programme shows the maximum usable frequency (MUF) and the lowest usable frequency (LUF), and at times when the LUF exceeds the MUF, HF skywave communication is not possible. The MUF is determined by the degree of layer ionisation whilst the LUF is generally determined by the multi-hop path attenuation and the noise level at the receive site. The frequency of optimum transmission (FOT) is the frequency with maximum availability and minimum path loss, and is generally taken as 90% of the MUF.

HF groundwave and skywave communications are still widely used for long distance low capacity services such as aeronautical ground to air transmissions, despite the increasing availability of satellite services. Defence forces are a major user, but the effects of the electromagnetic pulse (EMP) released by a nuclear explosion on the ionosphere and on equipment require special consideration in military systems. Much of the information on this topic is classified, but some papers are available.

2.7 Terrestrial line of site propagation

In the VHF, UHF and microwave bands, the ionosphere has a minimal effect on propagation although anomalous conditions such as sporadic E propagation do affect the lower frequencies in this range. The frequencies are generally well above the critical frequency so that transmissions pass through the ionised layers and out into space. This offers the considerable benefit that the same frequencies can be used and re-used many times without causing mutual interference, provided sensible frequency planning is carried out and adequate physical separation is provided.

Communications link calculations can be carried out using the Friis power transmission formula as in Equation 2.12. where P_r is received power in watts, P_t is transmitted power in watts, G_t is the gain of the transmitting antenna in the direction of the receiving antenna, G_r is the gain of the receiving antenna in the direction of the transmitting antenna, and r is the distance in metres between antennas.

$$P_r = \frac{P_t\, G_t\, G_r\, \lambda}{(4\,\pi\,r)^2} \tag{2.12}$$

For systems calculations, this can be expanded to include transmit and receive antenna VSWR and polarisation mismatch as in Equation 2.13, where ρ_r is the magnitude of the voltage reflection coefficient at the receive antenna, ρ_t is the magnitude of the voltage reflection coefficient at the transmit antenna, and p is the polarisation mismatch.

$$\rho_r = \frac{P_t\, G_t\, G_r\, \lambda^2\, (1 - \rho_r^2)\, (1 - \rho_t^2)\, p}{(4\,\pi\,r)^2} \tag{2.13}$$

Voltage reflection coefficient can be calculated from the VSWR as in Equation 2.14.

$$p = VSWR - \frac{1}{VSWR} + 1 \tag{2.14}$$

The mismatch between two elliptically polarised waves is given by Equation 2.15, where R_t is given by Equation 2.16 and R_r by Equation 2.17. R_t is the transmit antenna axial ratio, and R_r is the receive antenna axial ratio.

$$p = \frac{1 + R_t^2 R_r^2 + R_t R_r \cos 2\theta}{(1 + R_t^2)(1 + R_r^2)} \tag{2.15}$$

$$R_t = r_t + \frac{1}{r_t} - 1 \tag{2.16}$$

$$R_r = r_r + \frac{1}{r_r} - 1 \tag{2.17}$$

Equation 2.13 assumes free space conditions, however absorption occurs due to rain and fog and also water vapour and oxygen in the air.

Although VHF, UHF and microwave point to point communications is frequently referred to as 'line of sight', the change in the refractive index of the atmosphere with height does in fact cause radio waves to be bent in the same direction as the earth's curvature. This in effect extends line of sight, and the relationship between the surface refractivity and the effective earth's radius is shown in Figure 2.12.

The refractive index of air does in fact depend on atmospheric pressure, water vapour pressure and temperature and all these factors vary. An average effective earth radius factor of 1.33 or 'four thirds earth' is therefore assumed for most link assessments.

The profile of the topology between transmitting and receiving antenna can also be plotted. It is desirable to avoid any obstructions

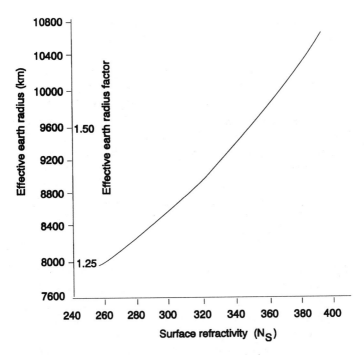

Figure 2.12 Atmospheric surface refractive index versus effective earth radius

within the first Fresnel zone which is defined as the surface of an ellipsoid of revolution with the transmitting and receiving antennas at the focal points in which the reflected wave has an indirect path half a wavelength longer than the direct paths. Figure 2.13 illustrates this and the radius of the first Fresnel zone at any point (P) is given by Equation 2.18, where R is the radius, and d_1 and d_2 are the distances from point P to the ends of the path.

$$R^2 = \frac{\lambda \, d_1 \, d_2}{d_1} + d_2 \tag{2.18}$$

The height of the transmitting and receiving antennas above the intervening terrain and the roughness of the terrain have a marked

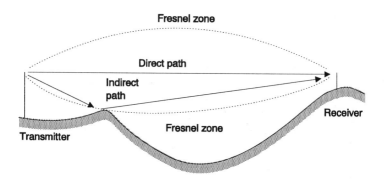

Figure 2.13   The Fresnel zone

effect on path attenuation and received field strength in the VHF and UHF bands. The Friis formula will generally yield results that are too optimistic i.e. attenuation is too low and field strength is too high. CCIR Recommendation 370 (CCIR, 1982) reproduced in Figure 2.14 shows received field strength values at 10 metres above ground level for a radiated power of 1kW for different transmit antenna heights. The field strengths are modified by correction factors for the terrain irregularity between transmit and receive antennas which can increase attenuation by up to 18dB or reduce it by 7dB depending upon the topology. Average terrain irregularity is defined as differences in height above and below 50 metres in only 10% of the path length. Figure 2.15 (CCIR, 1982) gives attenuation correction factors for field strength for various terrain height differences.

Frequencies at 900MHz and 1800MHz are used for comparatively short distance "cellular" services, and communications will often be limited by multi-path interference caused by single or multiple reflections, particularly in urban environments. The only practical solution is to try different positions for the transmit antenna to minimise this problem. Multi-path interference can also be a problem on microwave links for TV outside broadcasts where transmissions may be from moving vehicles or aircraft. This can be minimised by using circular polarisation, as the reflected signals will undergo a polarisa-

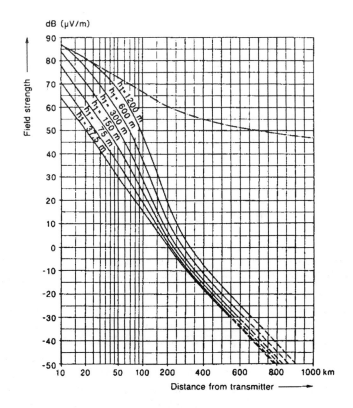

Figure 2.14 Received field strengths for different transmit antenna heights, v.h.f. and u.h.f. bands

tion reversal which will enable the receive antenna discriminate against them.

As the propagation medium is continuously changing, the figures derived from CCIR curves are based on statistical probabilities, generally 50% signal levels for 50% of the time, and therefore diversity transmission and a larger link margin may be required to increase channel availability.

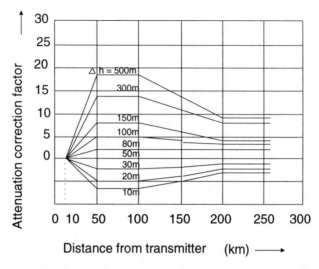

Figure 2.15 Attenuation correction factors for different terrain roughness

2.8 Over the horizon transmissions

The four thirds earth radius does extend transmission beyond the horizon in normal conditions, but a natural phenomenon known as ducting can extend this considerably. Ducting happens in stable weather systems when large changes in refractive index with height occur and cause propagation with very low attenuation over hundreds of kilometres. The propagation mode is similar to that in a waveguide. Ducting is not a very predictable form of propagation and therefore cannot be used commercially, but information is available in CCIR Report 718 (CCIR, 1982).

Tropospheric scatter and meteor scatter communications both provide over the horizon communications on a regular and predictable basis. Tropospheric scatter relies on small cells of turbulence in the troposphere between 2km and 5km above the earth surface which scatter incident radiation, as illustrated in Figure 2.16. Frequencies in the lower part of the microwave range from 1.0GHz to 5.0GHz are

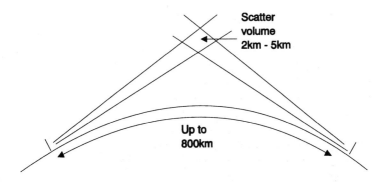

Figure 2.16   A tropospheric scatter communications link

generally used as this gives the maximum amount of forward scatter, nevertheless the amount of energy that is scattered forward is extremely small and the losses in the scatter volume are in the order of 50dB to 70dB in addition to the free space loss. Large short term variations following the Rayleigh law also occur and therefore quadruple or multiple diversity is almost always used. This is achieved by employing two antennas, horizontal and vertical polarisation, different scatter volumes (angle diversity) and different frequencies. Space and polarisation diversity generally yield the best results. Tropospheric scatter can provide communications up to 500km, but paths as long as 800 kilometres have been operated.

Meteor scatter communications makes use of ionised trails of meteors burning up on entering the earth's atmosphere. This is not such a rare event as might be thought, and there is a steady shower of material entering the earth's atmosphere although there are random and predictable variations. Meteors that leave usable trails have diameters between 0.2mm and 2mm, and the trails which last for around half a second to several seconds occur at a height of about 120km. They may extend up to 50km although 15km is a typical value. The number of particles entering the earth's atmosphere is inversely proportion to size, so larger meteors occur too infrequently to be of use whilst the numerous very small particles do not generate sufficient ionisation. The variation in the number of usable ionised trails occurs on a daily basis with more occurring during daylight as the earth's rotations tends to sweep up more meteors than at night. There are also predictable showers occurring throughout the year.

Meteor scatter communications can be established for paths from 200km to 2000km, but information has to be transmitted in bursts when a link is established so the medium cannot be used for services such as speech. Radio waves are reflected from the trails by different mechanisms depending upon the density of ionisation; however frequencies in the 30MHz to 100MHz range are most effective with 50MHz being the optimum frequency.

Horizontal polarisation is generally preferred. When a link is established, path attenuation is high due to the scattering process, and is typically 175dB for a 40MHz, 1000km path (Cannon, 1987). This is approximately 50dB greater than the free space attenuation of the same distance.

2.9 Propagation for satellite communications

Most communications satellites are placed in geo-stationary orbits 36000km above the equator, therefore transmitting and receiving earth stations can fix their antenna positions with only minor adjustments being required for small shifts in satellite position or changes in atmospheric propagation conditions.

Such orbits also have the advantage of providing potential coverage of almost one third of the earth's surface, but the disadvantage of high free space loss compared with lower non stationary orbits. The systems planner will need to calculate the link budget taking into account such factors as the satellite EIRP (equivalent isotropically radiated power) and receiver noise performance.

The free space loss can be calculated using Equation 2.7 and distance will have to take into account both the difference in latitude and longitude of the position of the satellite on the earth's surface to that of the transmitting or receiving station. If the distance and great circle bearing is calculated using Equations 2.10 and 2.11, the elevation and distance of the satellite can also be calculated. In addition to the free space loss, the loss due to atmospheric attenuation must be taken into account, and this will depend upon precipitation conditions in the earth station area.

Typical values at 11GHz would be 1.0dB for an 'average year' increasing to about 1.5dB for the worst month. The actual figures to be used should be calculated from local meteorological data and attenuation curves.

Calculation of satellite paths is not often required as operators usually publish 'footprint' maps showing the received power contours in dBW, taking into account the path loss and the radiation pattern of the satellite transmitting antenna.

2.10 Summary of radio wave propagation

2.10.1 VLF and LF: 10kHz to 300Khz

Long distance (greater than 1000km) ground wave transmission for communications and radio navigation. Propagation unaffected by ionospheric conditions. High power and large antennas required. Very limited spectrum availability.

2.10.2 MF: 300kHz to 3MHz

Medium to short distance (up to 1000km) ground wave transmission for sound broadcasting and mobile communications. Propagation little affected by ionospheric conditions, but night time interference can be a problem.

2.10.3 HF: 3MHz to 30MHz

World wide communications using comparatively low power (1kw), but heavily dependent on ionospheric conditions. Limited channel availability. Used for sound broadcasting, point to point and mobile maritime and aeronautical communications.

2.10.4 VHF and UHF: 30MHz to 1GHz

Officially, UHF extends to 3GHz. Lower frequencies are affected by anomalous propagation conditions. Line of sight communications, typically 80km (meteor scatter on 50MHz can provide services up to 2000km). Used for multi-channel point to point communications, FM sound and TV broadcasting and mobile communications.

2.10.5 Microwave: 1GHz to 30GHz

Unaffected by ionospheric conditions, but some attenuation at higher frequencies due to rain. Line of sight communications, typically 50km (tropospheric scatter can provide services up to 500km). Satellite services provide world wide coverage. Wide band multi-channel communication available. Extensively used for terrestrial point to point and satellite communications, also radar.

2.10.6 Millimetric: 30GHz to 100GHz

Unaffected by ionospheric conditions but moderate to severe attenuation due to atmospheric conditions. Limited line of sight communications, typically 10km. Limited usage, mainly short distance speech and data links but increasing usage is likely.

2.11 References

Boithias, Lucien (1987) *Radio wave propagation*, North Oxford Academic Publishers.

Braun, G. (1986) *Planning and engineering of short-wave links*, 2nd Ed. John Wiley & Sons, Chichester.

Budden, K.G. (1985) *The propagation of radio waves*, Cambridge University Press.

Cannon, P.S. (1987) The evolution of meteor burst communications systems. *J. IERE.* **57** (3) (May/June).

CCIR (1978) *Man-made radio noise*, Report 258- 3, ITU, Geneva.

CCIR (1982) *Volume V, Propagation in non-ionised media*, Geneva.

CCIR (1984) *World distribution and characteristics of radio noise*, Report 233-1, ITU Geneva.

Damboldt, T. (1975) A comparison between Deutsche Bundespost ionospheric h.f. propagation predictions and measured field strengths, *NATO AGARD, Radio systems and the ionosphere, Conf. Proc. 173*, Athens, (May).

Dick, M.I. (1990) *Propagation model of CCIR report 894,* Rutherford Appleton Laboratories, (May).

Friis, H.T. (1946) A note on a simple transmission formula, *IRE Proc.* (May), pp. 254-256.

Hitney, H.V. (1990) *IONOPROP, ionospheric propagation assessment software and documentation*, Artech House, London.

Johnson & Jasick (1984) *Antenna Engineering Handbook,* McGraw Hill.

Longmire, C.L. (1978) The electromagnetic pulse produced by nuclear explosions, *IEEE Transactions on Electromagnetic Compatibility,* **EMC-20** (1) (February).

Maslin, N.M. (1987) *H.F. Communications, A systems approach,* Pitman, London.

Millington, G. (1949) Ground wave propagation over inhomogeneous smooth earth, *J. Inst. Electr. Eng. Part 3.*

Picquenard, Arme (1984) *Radio wave propagation*, Philips Technical Library, McMillan Technical Press.

Radio Society of Great Britain, *Radio Communication Handbook.*

Stanniforth, J.A. (1972) *Microwave Transmission*, English University Press.

Stark, Axel (1986) *Propagation of Electromagnetic Waves*, Rohde & Schwarz.

3. Radio spectrum management

3.1 Introduction

Governments manage the use of the radio spectrum to prevent the degradation by interference of a valuable resource, to ensure that radio systems use spectrum economically, leaving room for new users and new uses, and to provide for the orderly use of frequencies where this is necessary for practical reasons.

To ensure that these objectives can be attained, governments insist that there should be no transmission of radio signals without permission, signified by a licence. Licences are also required for many receiving stations. Exceptionally, a general authorisation may be given for specified frequency bands to be used for specific kinds of very low power devices without formality, it being assumed that the possibility of interference can be disregarded in such cases.

Most licences specify the carrier frequency which has been assigned for use at a named station, together with other technical parameters of the emission (such as the transmitter power, the bandwidth that may be occupied and key antenna characteristics) and the purpose for which the system may be used. The assignment of a frequency usually implies an assurance that unacceptable interference is not likely to occur and that action will be taken by the licensing authority to eliminate such interference if it should occur. Typically a fee is charged for the licence, to cover the administrative costs of issuing it and a share of the cost of national spectrum management.

In pursuit of their policy towards industry, governments also regulate telecommunication systems and in particular, the operation of commercial systems set up to provide facilities for sale to other parties. Where the telecommunication systems use radio, the regulatory process may be implemented through the issuing, or withholding, of licences to operate radio stations. However, regulation of this kind is not discussed here.

If frequencies were assigned to stations at random from any technically suitable part of the spectrum, radio system development would be chaotic and the spectrum would be used inefficiently. The necessary degree of order is obtained by categorising radio systems into services and allocating specific frequency bands for each service. More detailed provisions are made for regulating the use made of assignments in some bands. The services and the allocations are outlined in Section 3.4 below.

Each government is sovereign in spectrum management within its country's frontiers. However, frontiers are no barrier to interference. Furthermore, operational radio equipment crosses frontiers on ships, on aircraft and by land and should desirably be usable wherever it goes. Governments collaborate where necessary in the management process to limit cross-frontier interference and to facilitate the operation of mobile radio stations. Much of this collaboration is done informally and bi-laterally or multi-laterally between the governments of neighbouring countries. Some specific management functions are co-ordinated through recognised international organisations with specialised responsibilities; for example, national use of some frequency bands allocated for civil aviation purposes is co-ordinated through the International Civil Aviation Organisation (ICAO). However, the International Telecommunication Union (ITU) is the primary world forum for radio spectrum management.

Governments set up bureaux to manage the spectrum. Usually such a bureau is part of a government department but some functions may be devolved to a body responsible to, but outside, government. The national authority, however it is constituted, is known as the administration.

3.2 Frequency assignment

3.2.1 National frequency assignment

Each administration maintains a register of the frequency assignments that it has made, recording where each assigned frequency is transmitted and received and the more important parameters of emissions. The use that is made of the spectrum is monitored, to ensure that unauthorised transmitting is stopped and that transmitting licences that are not put to use are withdrawn.

When a new assignment is to be made and the emission parameters required for satisfactory operation, such as the transmitter power and the bandwidth to be occupied, have been determined, a search can be made through the register to identify frequencies in appropriately allocated bands that may be suitable for assignment. The information in the register and predictions of propagation conditions are used to estimate whether any of these frequencies could be used for the new emission without causing unacceptable interference to other systems within the national jurisdiction. Similarly estimates are made of the interference that a new receiving station would suffer from established transmitting stations within the national jurisdiction if its receiver were tuned to a proposed new frequency. When an apparently interference free frequency has been found in this way, by inspection, it may be assigned for the new use, details being entered in the national register to prevent further assignments being made that would cause or suffer interference in the future.

Software is available for applying computer methods to the identification of interference free frequencies that could be used for new systems without causing interference to existing, registered, assignments. Spectrum efficient use of such techniques, however, requires the information in the register to be available in database form, together with adequate topographical and radio propagation information.

For some kinds of radio station, some administrations devolve the task of selecting frequencies for assignment to the owners of the stations. The relevant parts of the national register of frequency assignments and all proposals for new assignments are published and existing assignees, would-be assignees and the expert agents of both are free to argue before the administration whether a proposed new assignment should be made.

This description of the process of choosing frequencies for assignment by inspection is idealised, no mention being made of major practical problems. The most important of these problems is the international dimension, considered in Section 3.2.3. Other difficulties may arise within national frontiers. For example, the topographical and radio propagation data which are available may not be accurate enough to permit the wanted signal level and the interference levels to be determined with sufficient confidence, by calculation, to ensure that spectrum is used efficiently. The national register of frequency assignments may be inaccurate, leading to sub-optimal new assignments or interference when new assignments are taken into operation. Or there may not be a suitable frequency available in a heavily loaded frequency band without the re-arrangement of frequencies already assigned to other stations. Even without such complications, it is a labour intensive process demanding a lot of expertise. Furthermore, being a random process, it may not provide optimum use of the spectrum.

There are various ways of making this process more systematic, easier to apply and, in favourable circumstances, more efficient in using the spectrum. Thus:

1.   A national radio channelling plan is sometimes drawn up. For example, when planning for bi-directional links, a frequency band would be split into two equal sub-bands, A and B, and each sub-band would be sub-divided into channels, each wide enough for a single uni-directional link. See for example Figure 3.1. One channel from each sub-band would be assigned to each bi-directional link, as in Figure 3.2. The plan shown in Figure 3.1 is suitable, for example, for multiplexed telephony links carrying 24 channel FDM with FM or 30 channel PCM with 4 phase PSK. Such a plan can facilitate the selection of frequencies for assignment. Where the planned frequency band is to be used for a large number of stations with similar characteristics, a regular radio channelling plan may give other advantages also; for example, the use of frequency synthesisers to determine the operating frequencies of transmitters and receivers may be facilitated and efficient spectrum utilisation may be more readily achieved through the imposition of strict type approval specifications for transmitters and receivers.

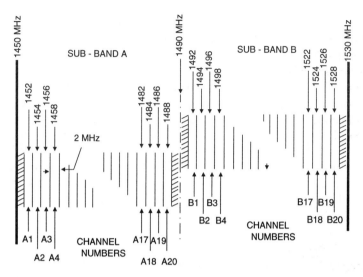

Figure 3.1 Radio channelling plan, providing channels 2MHz wide

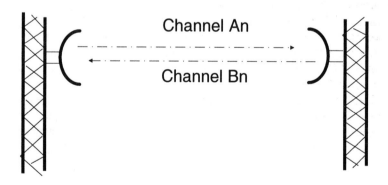

Figure 3.2 Channels An and Bn from the plan in Figure 3.1 are assigned to a bi-directional link

2.   The radio channelling plan concept might be extended to become a national frequency allotment plan. A wide geographical area, perhaps nation-wide in extent, would be divided into cells, and a list of channels would be allotted for assignment within each cell. The same channel would be allotted for assignment in other cells, with enough geographical separation to make the probable level of interference acceptably low. The plan might, in the simplest case, consist of a regular tessellation of hexagonal cells, each allotted an equal number of channels. Alternatively, an irregular plan based, for example, on administrative areas and channel allotments proportionate to the local level of demand, might be more convenient to manage and would not necessarily be less efficient in spectrum use.

3.   There will be circumstances in which the locations of stations that will need assignments in a frequency band in the future are known in advance. In such cases a frequency assignment plan, optimising the geographical dimension of spectrum utilisation, might be prepared at the outset, assignments being made in accordance with the plan as each station starts operating.

Other measures may also be implemented to reduce the burden of spectrum management that would otherwise fall on administrations. For example, where a single radio user has need of a great deal of spectrum, that user might be allotted a complete frequency band on condition that it is used efficiently.

Administrations may also license stations or operators without undertaking to control interference. For example, the radio amateurs' societies are expected to discipline the use of bands allocated to radio amateurs.

There are alternatives to this method and some of them are described in Sections 3.4 and 3.6. However, despite the difficulties that are involved, the basic process of assigning for a new requirement a frequency that has been found to be suitable by inspection of the national register of frequency assignments, aided where feasible by national spectrum planning as indicated above, is in very general use.

3.2.2 International registration of assignments

Many transmitting stations in small countries and most stations in large countries do not radiate enough power to interfere with receiving stations abroad, particularly in the higher frequency ranges. Many receiving stations are likewise unlikely to suffer interference from abroad. The national frequency assignment management arrangements described in Section 3.2.1 are adequate for these stations. However, interference can cross frontiers; some radio links are international and some radio equipment moves from country to country and is used wherever it goes. The international aspects of frequency assignment management fall into three parts; the international machinery for registering frequency assignments is reviewed in this section, the management of spectrum in an international context follows in Section 3.2.3 and the elimination of interference if it should occur is considered in Section 3.2.4.

Before an administration makes a frequency assignment, it should ascertain whether the activation of the proposed new assignment will involve interference with already established foreign assignments. This often involves consultation with other administrations. In some circumstances, and in particular where it may be difficult to identify the foreign administrations that should be consulted, there are agreed mandatory procedures governing the consultation process (see Section 3.2.3). However, informal contacts are used if consultation between the administrations of adjacent countries is likely to be sufficient. Then:

1. If an assignment, when activated, may be capable of causing interference abroad or if it is to be used for an international link, details of the assignment must be notified to the ITU Radiocommunications Bureau (ITU/RB).

2. The ITU/RB examines each notice to verify that the assignment is in accord with international agreements, and that any mandatory steps have been taken to ensure that interference will not occur to recognised foreign assignments. In some frequency bands below 28MHz, in which it is particularly difficult to assess whether interference is likely to occur, the

ITU/RB itself makes a technical examination of the interference prospects. If the newly-notified assignment survives these examinations, it is registered in the Master International Frequency Register (MIFR) and dated. The essence of the MIFR is published twice a year in microfiche by the ITU, as the International Frequency List (IFL). A list of newly notified national assignments, together with an up-dating of the IFL, is circulated weekly to administrations by the ITU/RB.

Priority of registration in the MIFR helps to establish international recognition of the right of a station to operate on the frequency nationally assigned to it; see Section 3.2.4.

3.2.3 Limitation of cross-frontier interference

The use of the inspection method for identifying frequencies for assignment, laborious when interference from abroad can be disregarded, becomes even more difficult when cross frontier interference must be considered. There are indeed circumstances when effective international consultation on frequency assignments is virtually impossible, because the parties with whom consultation might be necessary cannot be identified with certainty or are too numerous. Such problems are made even more complex in many frequency bands by the practice of allocating spectrum, not to one service, but to several; this is called frequency allocation sharing.

Furthermore, many administrations are dissatisfied with the principle whereby the right to use a frequency may be based solely on priority of international registration of a national assignment. They consider that this benefits developed countries with long established radio systems, to the disadvantage of developing countries which are in the process of setting up their infrastructure. The impact of the 'first come, first served' concept may be particularly serious where interference can arise at great distances, typically in the HF range, 3 to 30MHz, and also where satellite systems are involved.

Various special measures have been adopted in particular frequency bands to deal with these problems and objections. The most important of these measures are as follows:

1. International radio channelling plans have been adopted in a number of frequency bands. These plans specify the carrier frequencies that should be used and the kinds of traffic that they should be used for. In some cases they specify also various parameters of the emissions that may be employed.

2. International frequency allotment plans have been formally agreed for some frequency bands. These plans extend the principle of radio channelling plans, typically by allotting channels to administrations for assignment to stations within their jurisdiction. See also Section 3.6.2.

3. International frequency assignment plans have been formally agreed for certain other frequency bands. These plans specify virtually all of the details of the foreseen assignments in a frequency band, including the location of stations, as well as many of the technical parameters of the foreseen emissions. See also Section 3.6.2.

4. In situations where it is difficult to identify all of the administrations that might be affected by a new frequency assignment, informal liaison between administrations would not be satisfactory. Detailed mandatory international frequency co-ordination procedures have been agreed in such cases.

5. Sharing constraints, such as constraints on transmitted power, are imposed on stations to limit the interference that other stations will suffer to an acceptably low level, in cases where frequency co-ordination would be impracticable.

Applications of these various specific measures are reviewed in Section 3.4.

3.2.4 Elimination of interference

If all the necessary preliminary checks and consultations have been carried out effectively before a new assignment is brought into operation, interference should not arise. However, radio propagation predictions are imperfect, the use that will be made of frequency assignments is not always clearly foreseen at the outset and the need grows every year to squeeze more and more systems into the most desirable frequency bands. Interference does arise.

If the station suffering interference provides what is called a safety service, typically a communication or radio navigation system upon which the safety of human life may depend, the interfering station stops using the assignment immediately.

If the assignment suffering interference and the assignment causing interference have both been made by the same administration, then that administration resolves the problem, typically by assigning another frequency to one of the links.

If the frequency assignments causing and suffering interference fall within the scope of an internationally agreed frequency allotment or frequency assignment plan, the interference may have arisen because system parameters assumed in planning have not been implemented; if so, the remedy is to bring the systems into line with the plan, or to obtain the agreement of all parties concerned to revise the plan to eliminate the interference. However, if the assignment suffering interference is operating in accordance with a plan and the interfering assignment, having been made by another administration which has ratified the plan is, nevertheless, not included in the plan, then the interfering assignment will be withdrawn.

Few frequency bands are allocated internationally for only one service; most bands are shared by at least two services, and some by several; see Section 3.3.1. The services sharing a band may have equal allocation status, but in other bands some allocations have primary status, and others have secondary status. Thus, station 'S', of a service having secondary allocation status must stop using an assignment that causes harmful interference to station 'P', belonging to a service with primary status, and station S has no remedy if station P interferes with its signals.

If safety considerations do not arise, if there is no internationally agreed plan, and if the radio services to which the stations involved belong are the same, or have equal allocation status in the frequency band, the operating organisations involved, assisted if necessary by their administrations, determine together what could be done to eliminate the interference. This may involve modifying one or other of the stations or modifying the emissions. As a last resort, it is customary for the opportunity to use the frequency without interference to be conceded to the station with the earlier date of registration in the MIFR but administrations are not committed to doing this.

3.3 International frequency allocation

3.3.1 Allocations for radio systems

In the early years of radio, only the lowest few megahertz of the radio spectrum were in use and they were used for point-to-point communication between fixed stations, communication with and between mobile stations, mainly ships, and broadcasting. These applications were called the fixed service, the mobile service and the broadcasting service respectively. By international agreement the spectrum then in use was divided into several frequency bands, different bands being allocated for each service.

This concept of dividing the spectrum between the different kinds of radio service is still found to be wise and its application has been extended and elaborated to serve modern requirements. The international table of frequency allocations (ITU, 1990, Article 8) now covers the frequency range 9kHz to 275GHz, divided into hundreds of frequency bands, allocated for 33 different services.

Different countries need the various kinds of radio facilities in different proportions. Consequently, a rigid and uniform global pattern of frequency allocation would not be acceptable. The need for flexibility has been met in three ways:

1.   Many frequency bands have been allocated, not for one but for several services; these bands are said to be shared. Usually all services sharing a band have equal rights of access to it, but there are bands in which one service is given primary status and another has an inferior, secondary status. As mentioned in Section 3.2.4, the status of the allocation for a service affects the treatment that a station of that service receives if interference should arise.

2.   The international frequency allocations are not uniform world-wide. Three geographical regions have been defined

(ITU, 1990, para 392-399) and in some frequency bands there are different allocations in the different regions. The regions are approximately as follows:

Region 1: Europe, Africa, CIS in Asia, Mongolia and Asian countries to the west of the Persian Gulf.

Region 2: North and South America.

Region 3: The parts of Asia not in Region 1, plus Australasia.

3. Footnotes to the international table of frequency allocations provide for departures from the world wide or regional allocations in various specified countries. See Section 3.5 below.

The frequency bands allocated to the various radio services and the methods which are applied internationally by regulation to manage the assignment of spectrum in the various bands are reviewed in Section 3.4 below.

Amendments to the international table of frequency allocations and the Radio Regulations are discussed and agreed at World Radio Conferences of the ITU (WRC). WRCs were known as World Administrative Radio Conferences (WARCs) before 1993.

3.3.2 Allocations for industrial, scientific and medical applications

Not all generators of radio frequency energy are radio systems. Some industries use high frequency power, for example, for the heat treatment of metals and for drying and welding non-metallic materials. Some scientific machines, most notably sub-atomic particle accelerators, use radio frequency energy on a massive scale. Some apparatus used in medical diagnosis and treatment also uses radio frequency energy. Domestic and commercial microwave ovens are everywhere. All of these devices radiate some energy inadvertently. Similarly there is substantial radio frequency radiation from electrical power transmission plant, electrical traction systems and above all from petrol driven motor vehicles.

Often the radiation from such equipment is wide band and noise like but there may be a dominant spectral component. The designers and users of such equipment are urged to minimise both kinds of radiation. In addition, however, a dozen frequency bands spanning the spectrum from 6MHz to 250GHz have been identified by the ITU in collaboration with the International Electrotechnical Commission (IEC), to be used for Industrial, Scientific and Medical (ISM) equipment which has a narrow band radiation characteristic. These bands are also allocated for radio use, most commonly the fixed or the radiolocation services, it being understood that interference from ISM equipment must be tolerated in these bands.

3.4 International spectrum

3.4.1 Overview

Almost all radio systems which provide telecommunications facilities fall into one of two broad categories, namely systems that link fixed stations and those providing links to, and between, mobile stations. These two categories, including the satellite systems for each, comprise the fixed services and the mobile services. Extensive frequency bands have been allocated for them, and many are shared between them. However, there are major differences in the regulatory methods used in the international management of frequency assignments to fixed and mobile stations and spectrum management for the two groups is considered separately in Sections 3.4.2 and 3.4.3. The amateur services follow in Section 3.4.4. The inter-satellite service, providing spectrum for direct satellite links to systems for any radio service, is reviewed in Section 3.4.5.

However, there are other kinds of radio system. About one half of the spectrum below 1GHz is allocated for broadcasting. About one third of the spectrum between 1GHz and 20GHz is allocated for radar and other position determination systems. There are lesser allocations for other specialised applications for radio. Some of this spectrum is shared with radio systems used for telecommunications and the conditions arising from sharing have significant impacts on the telecommunications services. There are brief accounts of these allocations for non-telecommunications radio services, as they affect telecommunications systems, in Sections 3.4.6 to 3.4.8.

3.4.2 Fixed station links

3.4.2.1 *Terrestrial links*

There are extensive allocations for the fixed service, comprising point to point and point to multipoint radio links, in most parts of the radio spectrum.

The fixed service allocations below 150kHz, at one time very important, are now little used. Those bands are occupied nowadays mainly by radionavigation systems.

About 55% of the bandwidth of the HF spectrum, in about 40 narrow bands between 4MHz and 28MHz, is allocated exclusively for the fixed service and is in substantial use, mainly for long distance ionospherically propagated links.

Substantial bandwidth is allocated for the fixed service between 28MHz and 1GHz. However, these bands are shared with broadcasting and the mobile services. The fixed service has little opportunity to use these bands in areas, like Europe, where there is heavy demand for these other services.

Very wide bands are allocated for the fixed service above 1GHz, much of this bandwidth being shared with the mobile service, the fixed-satellite service or others. The means that have been adopted to deal with inter-service interference arising from this sharing pattern are considered in Sections 3.4.2.3 to 3.4.2.5. These allocations are heavily used, mostly for radio relay systems.

The Radiocommunication Assemblies of the ITU (ITU-R) recommend radio channelling plans for microwave systems but in most other respects frequencies are chosen for assignment in the fixed service by inspection, with informal consultation between the administrations of neighbouring countries where necessary to ensure that interference is not likely to occur. After consultation, the responsible authority notifies the assignment to the ITU/RB, and the Bureau registers the assignment in the MIFR, provided that the assignments are in accordance with international agreements and subject below 28MHz to a technical examination to confirm that interference is unlikely to arise. However, if the station lies within the co-ordination area of an earth station operating in the same frequency band, the

ITU/RB will not register a new assignment until its use has been co-ordinated with the earth station. Similar limitations apply in frequency bands around 12, 14.6 and 17.7GHz that have been planned for the broadcasting satellite service and the associated feeder links (see Section 3.4.6).

3.4.2.2 *Links via satellites*

Wide microwave and millimetre wave frequency bands are allocated for the fixed-satellite service. These allocations are used for links between fixed earth stations via satellites.

Feeder links may also be assigned frequencies in these bands. A feeder link connects an earth station at a given location with a satellite which is not of the fixed-satellite service. The term covers, for example, links between coast earth stations and satellites serving ships, links feeding programme material to broadcasting satellites and links used for down loading data from a data collection satellite.

Some fixed-satellite bands are allocated for transmission in the Earth to space direction (up links). The others are for the space to Earth direction (down links). The important up link allocations are around 6, 7, 8, 13, 14 and 30GHz. The corresponding down link allocations are around 4, 5, 7.5, 11, 12 and 20GHz. Three further up link bands, around 11GHz (Region 1 only), 14.6GHz and 17.7GHz, have been allocated for the fixed-satellite service but reserved for feeder links to broadcasting satellites. Almost all of these are shared with the fixed service and the mobile service. Some allocations are also shared with other satellite services, the directions of transmission being opposite to those of the fixed-satellite service in some cases. With few exceptions, these allocations have primary status.

The modes of interference arising between fixed-satellite stations and terrestrial stations are illustrated in Figure 3.3. The corresponding interference modes from one satellite network to another are illustrated in Figures 3.4 to 3.6. Special methods have to be used to prevent such interference, because it can occur over great distances and very large numbers of terrestrial stations operating in the same frequency band may be in the field of view of a satellite.

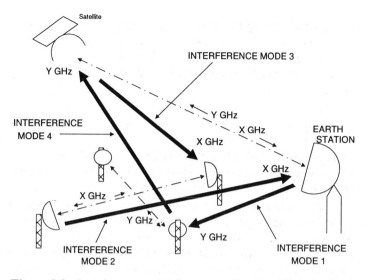

Figure 3.3 Interference modes between a fixed-satellite network and terrestrial links. (Wanted signal paths are shown in broken line)

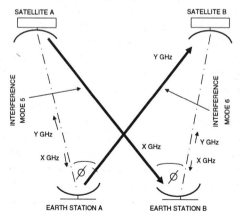

Figure 3.4 Interference modes from satellite network A to satellite network B, both using frequency band X for down links and band Y for up links. Interference from network B will enter network A by similar modes. (Wanted signal paths are shown in broken line)

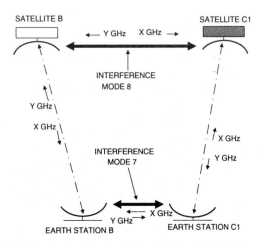

Figure 3.5 Interference modes between fixed-satellite networks having neighbouring satellites, both networks using frequency bands X and Y but in opposite directions of transmission. (Wanted signal paths are shown in broken line)

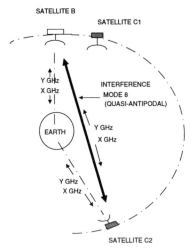

Figure 3.6 With frequency usage as in Figure 3.5, interference may also arise when the satellites are very wide apart in orbit, the quasi-antipodal case, as shown here

3.4.2.3 *Constraints on satellite transmitters*

Interference from satellite transmitters may enter fixed service receivers; this is interference mode 3 in Figure 3.3. It would not be feasible to control the level of this interference by negotiation, because of the very large numbers of stations that may be involved.

Most terrestrial stations are particularly sensitive to interference from satellites which are close to the horizon, that is, from energy which the satellite radiates towards the edge of the Earth's disc. See Figure 3.7. More powerful signals can be tolerated from sources higher in the sky. It has been estimated that interference would not exceed tolerable limits for terrestrial stations operating at 4GHz if the spectral power flux density from any one satellite, reaching the Earth's surface at an angle of elevation below 5°, does not exceed minus 152dB relative to 1 watt/m^2 in a sampling bandwidth of 4kHz. At angles of elevation greater than 25°, a flux density 10dB higher would be acceptable, with a linear dB transition between 5° and 25°.

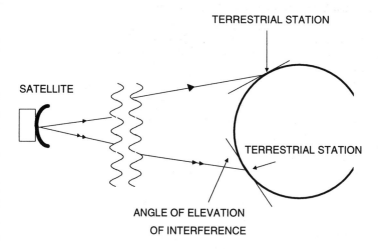

Figure 3.7   Geometry of interference from satellite transmitter to terrestrial service receiving station, showing angle of elevation of interference on arrival

Accordingly, these values have been adopted as a mandatory sharing constraint upon the fixed-satellite service at 4GHz wherever the frequency band is also allocated internationally, with primary status, for the fixed service; see Figure 3.8. Slightly less stringent constraints have been agreed for down link bands at higher frequencies.

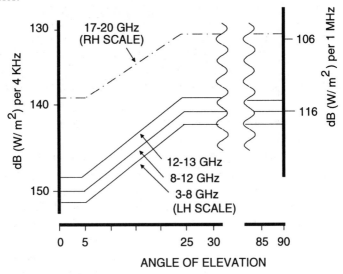

Figure 3.8   Constraints on the spectral power flux density at the Earth's surface from satellite transmitters operating in frequency bands shared with terrestrial services, as a function of δ and the frequency band

3.4.2.4   *Constraints on terrestrial transmitters*

Constraints on the power of fixed service transmitters operating in bands below 30GHz allocated for fixed satellite service up links (interference mode 4 in Figure 3.3) have been adopted to limit interference to satellite receivers for the same reasons as led to power flux density limits for satellite down links. The constraints are as follows:

1.   The power input to the transmitting antenna shall not exceed +13dBW below 10GHz nor +10dBW above 10GHz.
2.   The maximum equivalent isotropically radiated power (e.i.r.p.) shall not exceed 55dBW.

Below 10GHz the constraint is made more stringent in the direction of the geostationary satellite orbit. Thus, if the beam of a fixed service transmitter is directed within 0.5^o of the orbit, usually because it is pointed towards one of the two azimuths at which the line of the orbit cuts the local horizon, the maximum e.i.r.p. is 47dBW, rising in a dB linear mode to 55dBW in directions 1.5^o away from the orbit.

3.4.2.5 *Frequency co-ordination around earth stations*

There are agreed limits on the power radiated from earth stations at low angles of elevation in frequency bands shared with the fixed service. These limits are a sharing constraint, designed to keep within reasonable bounds the distances at which interference to terrestrial stations can occur. However, the limits are high, and they are unlikely to have much impact on the design of modern satellite networks.

The methods used in managing microwave spectrum for the fixed service, in the absence of earth sations, assume that the incidence of interference will be relatively predictable and interference distances will be relatively short. Thus, administrations discuss proposed new assignments with their immediate neighbours, if it is thought necessary, before notifying the assignments to the ITU/RB for international registration, but this action is informal and non-mandatory and international registration is not always sought if interference to or from future foreign terrestrial stations is thought to be unlikely.

However, earth stations have more powerful emissions than typical terrestrial stations and more sensitive receivers; interference may arise at great distances and such interference may be sporadic, depending, for example, on tropospheric duct propagation. A thorough and mandatory procedure is used to identify and eliminate potential cross frontier interference problems when an earth station is first set up. This procedure, once carried out, also provides a basis for co-ordination of subsequent significant changes of frequency assignments to both services in the vicinity of this earth station.

The procedure (ITU, 1990, Articles 11-13) is simple in principle, although somewhat complex in practice. The administration responsible for a proposed new earth station prepares a map, using agreed standardised parameters and techniques, showing the area around the proposed site within which, under worst case conditions, the earth station transmitter might cause significant interference to a receiver of the fixed service operating in the up link frequency band. A second map is drawn, showing the area within which, under worst case conditions, a transmitter of the fixed service operating in the down link frequency band might cause significant interference to the earth station receiver. These are interference modes 1 and 2 respectively in Figure 3.3.

If the co-ordination areas shown by these maps include any foreign territory where there is an agreed allocation of the same frequency bands for the fixed service, the two administrations exchange information on all relevant earth station and terrestrial station frequency assignments and co-operate to determine whether interference will exceed internationally agreed threshold levels. If interference problems are foreseen, acceptable solutions must be found before assignments are notified for international registration.

3.4.2.6 *Frequency co-ordination between satellite neworks*

A mandatory procedure is used to bring together for consultation the various administrations responsible for satellite systems that might interfere with one another. These consultations are often technically complex, since there are likely to be two bi-lateral interference modes (modes 5 and 6 in Figure 3.4) between any pair of satellite networks. There may be three other bi-lateral interference modes (modes 7 and 8 in Figures 3.5 and 3.6, including the "quasi antipodal" form of mode 8) in the circumstances, so far relatively rare, in which a band used for up links in one network is used for down links in the other, and vice versa.

An administration responsible for a proposed new satellite network publishes the essential technical details up to six years before the launch. This is called "advance publication". It draws the attention of other administrations, responsible for existing satellite networks, to possible future inter-network interference problems, enabling discussions to be held to seek ways of reducing prospective interference before the design and manufacture of the new satellite has progressed too far for changes to be made.

Special conditions apply to non-geostationary satellites operating in certain frequency bands below 3GHz; see Section 3.4.3.2. Elsewhere in the spectrum, for a satellite that is not to be geostationary, the frequency assignments for the receivers and transmitters on board the satellite can be notified to the ITU/RB for registration in the MIFR when these discussions have been completed, without further formality. In operation, the transmitters on a non-geostationary satellite and at the associated earth stations must be switched off whenever this is necessary to avoid causing interference to a geostationary satellite network.

However, for a geostationary satellite network, formal procedures follow the advanced publication phase to ensure that interference to or from other geostationary satellite networks will not exceed agreed acceptable limits. These procedures must be completed, to the satisfaction of the ITU/RB, before the satellite frequency assignments can be registered in the MIFR. Two different kinds of procedure are used for this purpose, according to the frequency bands in which the new satellite is to operate.

A frequency and orbit allotment plan (ITU, 1990, Appendix 30B) has been drawn up for geostationary satellite networks of the fixed-satellite service operating in the following frequency bands:

1. Up links at 6.725 to 7.025GHz and/or 12.75 to 13.25GHz.
2. Down links at 4.5 to 4.8GHz, 10.7 to 10.95GHz and/or 11.2 to 11.45GHz.

This plan provides for each country to set up one satellite network covering its own territory, subject to broad limits on the parameters of the equipment and the emissions that are used, whenever it is ready to do so, without risk that networks set up by other countries will have pre-empted its option.

In all other frequency bands there is a detailed procedure of frequency co-ordination, designed to give every administration with a reasonable concern about interference the right to discuss prospective interference levels with the administration proposing to set up the new network (ITU, 1990, Articles 11–13). The chief parameters that are open to negotiation when excessive interference is foreseen are the orbital locations of the satellites, frequency assignments and earth station antenna characteristics.

As will be seen from Section 3.4.6. frequency assignment plans have been agreed for satellite broadcasting and the associated feeder links in several frequency bands between 11.7GHz and 18.3GHz, some of which are shared with the fixed-satellite service. Special frequency co-ordination processes (ITU, 1990, Appendices 30 and 30A) additional to those outlined above, are used to ensure that broadcasting suffers no significant interference from the fixed-satellite service.

3.4.3 Links for mobile stations

3.4.3.1 *Terrestrial links*

Frequency bands are allocated for the mobile service, to be used for links with, or between, any kind of mobile station, but other allocations are made specifically for the aeronautical mobile, maritime mobile and land mobile services respectively. There is a further sub-division of allocations for aeronautical use as between 'Route' bands reserved for air traffic control purposes on civil air routes, and 'Off route' bands, used for other aeronautical purposes. Narrow bands at various frequencies have been identified for use in distress and other emergencies; special measures are taken to prevent interference arising in these bands.

There are maritime mobile allocations below 160kHz but their use is declining. The maritime band around 500kHz, limited to radiotelegraphy, is in heavy use. Maritime and aeronautical allocations around 2MHz and 3MHz are also in substantial use, although the mobile services must share these bands with other services.

Over fifty narrow bands, mostly unshared, spread between 4MHz and 28MHz, have been allocated for the maritime service or the aeronautical service (Route or Off route) and a few bands have been allocated in this part of the spectrum for the land mobile service. These bands are used mainly for long distance, ionospherically propagated links.

Most of the maritime bands have been sub divided between radiotelephony and radiotelegraphy, elaborate radio channelling plans have been agreed for the radiotelegraph sub bands and there is a frequency allotment plan for the radiotelephone sub bands. Most of the aeronautical bands have been incorporated into frequency allotment plans for Off-Route use (ITU, 1992a, Appendix 26 and Resolution 411) and Route use (ITU, 1983), suitable for radiotelephony but available for other kinds of signal also.There are important aeronautical mobile (Route) and maritime allocations around 125MHz and 160MHz respectively.

There is a radio channelling plan for the aeronautical mobile band and the use of these channels is managed mainly by the national civil aviation authorities under the aegis of ICAO. There is also a radio channelling plan for the maritime band (ITU, 1990, Appendix 18).

Wide bands between 28MHz and 470MHz and around 900MHz are allocated for the mobile services in general, shared mostly with broadcasting. These bands are used considerably for land mobile and aeronautical (Off route) systems but broadcasting is the major use. Since 1990 there has been increasing use of bands around 2GHz for land mobile systems.

Large amounts of spectrum are allocated for mobile services above 1GHz, sharing with the fixed service (and in many cases the fixed satellite service).

For technical reasons these allocations become less suitable for communication with moving vehicles as the frequency rises but considerable use is made of them for transportable equipment, often used for television outside broadcasting links and systems ancillary to the production of broadcasting programmes. Where these bands are shared with up links of the fixed satellite service there are power constraints on mobile service transmitters similar to those which govern the fixed service sharing the same bands.

3.4.3.2 *Links via satellite*

As with the (terrestrial) mobile services, frequency bands are allocated for the mobile-satellite service, usable for facilities for any kind of mobile station, but other allocations have been made specifically for the aeronautical, maritime and land mobile-satellite services. Some aeronautical mobile-satellite allocations are specifically reserved for 'Route' purposes.

The most important of the mobile-satellite allocations are between 1530MHz and 1660MHz; sub bands in this part of the spectrum have been allocated for maritime, aeronautical (Route) and land mobile purposes. There is little sharing with other services. These bands are used entirely for links between satellites and mobile earth stations, any associated feeder links between satellites and earth stations at permanent locations on the ground being assigned frequencies in bands allocated to the fixed satellite service; see Section 3.4.2. There are also frequency allocations to the mobile-satellite service in many other parts of the spectrum, many of them shared with the fixed-satellite service.

Several very narrow frequency bands below 1GHz and several wider bands between 1.5 and 2.7GHz were newly allocated to the mobile-satellite service in 1992 (ITU, 1992a), mainly to use with satellites in low Earth orbits (LEO).

3.4.3.3 *Spectrum management*

Very simple procedures are used for international registration when frequencies are assigned to stations at fixed locations (typically coast stations and aeronautical stations) in accordance with a frequency assignment or frequency allotment plan. In other bands, frequencies are chosen by inspection or from radio channelling plans, national or international, for assignment to coast stations, aeronautical stations, the base stations that communicate with land vehicles and the corresponding earth stations, typically after consultation with the administrations of neighbouring countries. These assignments are notified to the ITU/RB for registration after any necessary frequency co-ordination, as with the fixed and fixed-satellite services in corresponding ranges of frequency. The frequency co-ordination process for frequencies assigned to the satellites of mobile-satellite systems is as described for the fixed-satellite service in Section 3.4.2, exceptionally, an interim new procedure has been agreed for co-ordination in the new LEO bands (ITU, 1992a, Resolution 46).

Many kinds of mobile station are authorised to use, not one frequency, but any of many frequencies, the assignment used on any occasion depending on the circumstances. Assignments to mobile stations for terrestrial mobile services are not notified to the ITU/RB. Assignments to mobile earth stations, on ships, aircraft and land vehicles, are not notified individually for registration, but a collective notification is made by the responsible administration, indicating the area in which the mobile earth stations operate.

In choosing frequencies for assignment to mobile stations, an administration must often take care to exclude from the vicinity of, for example, a radio astronomy observatory, the use of a frequency at which the observatory operates.

3.4.4 The amateur services

There are allocations for the amateur service in every part of the radio spectrum in use today. Many of these bands are allocated for the amateur-satellite service also. However, many of these allocations are also shared with other services, typically the radiolocation service, and in some bands the amateur services have secondary status.

There are various internationally agreed technical constraints and operational constraints on amateur stations in the various bands. These constraints typically limit the power and bandwidth of emissions. Administrations authorising the launch of an amateur satellite operating in bands shared with other services undertake to ensure that the ground telecommand facilities will be good enough to keep interference from the satellite under control. The regulation of the amateur services differs from that of other services in that the licence is awarded, not to the station, but to the operator, subject to proof of competence to operate a station. Formal assignments are made to amateur satellites and notified to the IFRB. Specific frequency assignments are not made to other amateur stations.

3.4.5 The inter-satellite service

Some research services have frequency allocations for satellite to satellite links but the other satellite using services, which may find need for such links in the future, have no suitable spectrum allocated for them. Instead, frequency bands have been allocated for the inter-satellite service and frequencies in these bands may be assigned for any satellite to satellite links, regardless of the service to which the satellites belong.

The inter-satellite service allocations are all above 20GHz and most of them are in parts of the millimetre wave frequency range where absorption is high, due to molecular resonances in the atmospheric gases. All the bands are shared with other services but the inverse distance squared attenuation on potential interference paths, augmented by atmospheric absorption where it arises, makes troublesome interference to or from stations of other services improbable.

A frequency co-ordination process, based on the one referred to in Section 3.4.2.6 above, is used to limit interference between inter satellite links.

3.4.6 The broadcasting services

3.4.6.1 *Terrestrial broadcasting*

In Region 1 there is a broadcasting allocation around 200kHz. World wide there is another allocation within the approximate limits of 530kHz and 1700kHz. Several relatively narrow bands between 2MHz and 5MHz are allocated for broadcasting, mainly in tropical areas. There are extensive broadcasting allocations world wide between 4MHz and 26MHz. Almost all of these allocations are exclusive to broadcasting and there are internationally agreed frequency assignment plans for the bands below 1700kHz.

There are broadcasting allocations around 60, 100, 200 and 500 to 900MHz, used for television and FM sound radio. The broadcasting service has primary status throughout but there is extensive sharing with the fixed and the mobile services. However, access to spectrum for these sharing services differs considerably from place to place. In

Region 1 international frequency assignment plans have been agreed for broadcasting in all these bands and broadcasting use is intensive; substantial use is made of the mobile service allocations for land mobile systems, but only where this has been harmonised with the superior claim of the planned broadcasting assignments in nearby foreign countries. There are, however, no frequency assignment plans in Regions 2 and 3. In Region 2 the broadcasting service is protected from interference from the fixed and mobile services mostly by the secondary allocation status which the latter services have. In Region 3 all of these services have primary status and the partition of the bands between these various services is largely at the discretion of each administration, in consultation where necessary with neighbouring administrations.

Finally there are little used broadcasting allocations at 12, 42 and 85GHz. These bands are mostly shared with the fixed, mobile and broadcasting-satellite services. It is likely that the availability of these bands for the telecommunication services will be determined mainly by broadcasting satellite developments.

3.4.6.2 *Satellite broadcasting*

World wide frequency assignment plans have been agreed for the broadcasting satellite service (ITU, 1990, Appendices 30 and 30A) and for feeder links carrying programme material up to the satellites in the frequency bands given in Table 3.1. These plans have been designed for television broadcasting. There are other broadcasting-satellite allocations higher in the spectrum.

In the down link bands, which are shared with primary allocations for the fixed service and terrestrial broadcasting and primary allocations (secondary in Region 1) for mobile services, these sharing allocations may be used only in ways that cause no significant interference to satellite broadcasting which is in accordance with the plans, regardless of the date when satellite broadcasting starts. Similar protection of the planned feeder link assignments is also provided by the terms of the plans.

These planned frequency assignments, intended for direct broadcasting by satellite (DBS), are little used at present. However, televi-

Table 3.1   Frequency assignment plans for broadcasting satellite service

	Region 1	*Region 2*	*Region 3*
Broadcasting down link bands (GHz)	11.7 – 12.5	12.2 – 12.7	11.7 – 12.2
Feeder up link bands (GHz)	14.5 – 14.8 and 17.3 – 18.1	17.3 – 17.8	14.5 – 14.8 and 17.3 – 18.1

sion transmissions in the frequency bands allocated to the fixed-satellite service are received in millions of homes, constituting direct to home (DTH) satellite television.

A need has been perceived for sound broadcasting by satellite, powerful enough for reception by simple installations in motor cars. A frequency band has been allocated specifically for this purpose at 1452–1492MHz (ITU, 1992a, Article 8).

3.4.7 Location by radio

The radiodetermination and radiodetermination-satellite services include all systems which use radio to determine position, velocity or similar information, using terrestrial and satellite techniques respectively. Additional services have been defined to enable radionavigation systems, which provide safety services to be distinguished from radiolocation systems, which do not provide safety services, and to enable bands reserved for maritime systems to be distinguished from bands reserved for aeronautical systems.

With few exceptions the radionavigation service allocations are not shared with allocations to telecommunications services. Where such sharing does arise, radionavigation service assignments have advantages over those of the telecommunications services, if there should be interference, because the former always have primary allocation status. There are frequency assignment plans for radio navigation systems in some bands and additional protection arises from the 'safety service' status of the radionavigation service. Radiolocation service allocations are sometimes shared with allocations for telecommunications services.

3.4.8 Exploration and research using radio

The last group of services consists of the space research, earth exploration satellite, meteorological satellite, radio astronomy, space operation, meteorological aids and the standard frequency and time signal services, terrestrial and satellite.

Several of these services need frequency bands suitably located in the spectrum, free from man made signals, in which radio frequency energy radiated naturally from the earth, from the atmosphere or from cosmic sources can be measured. Several need suitably located frequency bands in which the earth, including its atmosphere, can be probed for information from satellites, as by radar, using 'active sensors'. And most of them need frequency bands for the transmission of acquired data and for satellite management.

For passive observations there are a few allocations below 50MHz, and about 50 other allocations, wide or narrow, mostly above 1GHz. Some of these allocations are exclusive to the passive services but others are shared with telecommunications services; administrations do what they can, through their frequency assignment practices, to protect radio astronomy observatories from interference. For active sensors there are 11 allocations between 1GHz and 80GHz, all shared with terrestrial radar. For data links the earth exploration satellite and meteorological satellite services each have one or two allocations between 1GHz and 20GHz, wideband and shared on equal terms with telecommunications services. The other services have in aggregate a substantial number of allocations in many parts of the spectrum for communication, but most are narrow, many have secondary status and all are shared with telecommunications services.

3.5 National frequency management

Governments, as members of the ITU, do not bind themselves to implement in all respects the world-wide or Region-wide provisions of the international table of frequency allocations. For example, many countries, claiming special needs, have obtained international agreement to the implementation of non-standard allocations within their territory; these agreements are recorded as footnotes to the international table. An administration is free to refrain from implementing within its jurisdiction any internationally agreed allocation. Also, an administration may assign frequencies for transmission in almost any frequency band to any kind of station, provided that the emission is not capable of causing interference abroad. Finally Article 48 of the Constitution of the ITU (ITU, 1992b) allows governments complete freedom of action with regard to military radio installations.

However, if use of an assignment contrary to the international table does cause harmful interference at a foreign receiving station which is operating in accordance with the table, then the interfering transmission must be stopped immediately.

Put briefly, governments do not bind themselves to implement the table, but they undertake to respect the right of other governments to do so.

Where a band is allocated internationally for several sharing services, it will often be desirable for an administration to limit the use of the band, or part of it, within its jurisdiction to some or only one of those services. For example, much of the spectrum between 47MHz and 960MHz is allocated internationally to both the broadcasting and the mobile services; in practice, within a given wide area, a given block of spectrum can be used for only one of these services. Likewise it is unsatisfactory for the same radio channel to be used by both aircraft and surface vehicles. Each country must decide for itself how to divide these bands between the various potential users. But, as indicated above, departures from the international pattern of frequency allocations lead to inefficient spectrum use and are to be avoided wherever possible.

Thus, an administration is free to construct a national frequency allocation table to meet its national requirements; see for example (HMSO, 1985). However, it is essential for the national table to be closely related to the international table and wherever possible to the national tables of neighbouring countries. In recent years European countries have aimed to standardise the bands used for mobile radiotelephone and paging systems and work is going on to extend standardisation to all bands by the year 2008.

To allocate frequency bands to services is only the beginning of the process of planning how the spectrum is to be used within the jurisdiction of an administration. Thus:

1. It will probably be desirable to designate solely for military use some of the bands allocated to the fixed, mobile, and radiodetermination services and the corresponding satellite services.

2. It may be found expedient to allot whole bands, or sub-bands, to major civil users of radio. Thus the recognised providers of public telecommunications services may be authorised to use, and be required to use efficiently, specified bands allocated to the fixed service. Other fixed service bands would be managed by the administration to provide frequency assignments for private fixed links.

3. Some bands allocated nationally for the land mobile service may be designated for private mobile radio systems; others would be designated for radio telephones giving access to the public telephone network, for communication with police and emergency service vehicles, for use by major public utilities, for cordless telephones, for wide area pagers, for citizens' band and so on.

Thus, the national frequency allocation table becomes an important and complex national planning document.

The uses of radio change with time, often because desirable new facilities have become technically and economically feasible and other, long established facilities have been superseded. It may be possible to provide for growth by assigning frequencies from frequency bands which are already in use, although it may first be necessary to reorganise the existing use so as to increase spectrum utilisation efficiency. Failing this, the usual alternative is to develop the use of vacant frequency bands, already appropriately allocated, high in the currently used spectrum.

However, in some countries most bands below 20GHz are already filling up and much higher bands are coming into use. Millimetre wave assignments are quite suitable for some kinds of radio system, such as short distance wide band fixed links, and they are acceptable for many other purposes, but they are unsuitable for, for example, most land mobile applications.

The emergence, in the 1980s, of technical means for providing convenient mobile radiotelephone facilities, car borne and hand portable, at an acceptable price has led to a massive demand for land mobile systems, cellular and trunked, and consequent pressure for spectrum below 1GHz.

To clear spectrum for these mobile systems, it has been necessary to replace many assignments for fixed links below 1GHz by new assignments at higher frequencies and a critical examination has been made of the need for some of the VHF and UHF allocations for broadcasting.

If provision cannot be made for a new allocation in a national table from amongst the options provided already by the international table, it will be desirable to get the international table amended at a competent World Radio Conference (WRC).

If there is general agreement that the addition of the new allocation to the international table is desirable, that allocation may be written into the international table of frequency allocations with world-wide, or perhaps Region-wide, applicability and with primary status. Often the already existing allocation will be retained for existing systems, creating a new sharing situation or extending one that was already in existence.

However, the need for changes of the international allocations may not be widely perceived and there may not be agreement as to what change should be made amongst the nations that do perceive the need. Failing agreement to a worldwide or regional primary allocation, international recognition of an allocation with sufficient status may be obtained by means of a secondary allocation, or a geographically limited footnote to the international table or by some combination of these measures by which the rights of existing stations of other services are protected.

3.6 Efficient spectrum utilisation

3.6.1 General measures

There are wise general principles of spectrum management, built into the ITU Radio Regulations, supported by the recommendations of the ITU-R, formerly the International Radio Consultative Committee (CCIR) and applied by administrations everywhere. These principles enhance the value of radio as a communications medium, increasing the information transmission capacity of the spectrum as a whole and minimising link degradations due to interference.

Attention is given to certain broad technical principles of spectrum conserving system engineering. There are international standards for the suppression of spurious emissions. The bandwidth occupied by emissions is required to be limited to what is necessary. Frequency tolerances are applied to radio carriers. Users of radio are discouraged from transmitting unnecessarily high power. Encouragement is given to the use of well designed receivers and antennas, capable of making the best of a medium which is interference limited in some frequency bands and some locations. These general requirements are undemanding, but more stringent measures are applied in particular situations and specific frequency bands where the need is evident; some of these latter cases are mentioned in Sections 3.6.2 to 3.6.4.

Above all, there is close collaboration between administrations at two main levels. In the absence of overriding political inhibitions, they co-operate to ensure that their frequency assignments do not cause or suffer interference. And they share expertise, in the technology of telecommunications and also in developing a precise statistical basis for predicting radio propagation conditions for wanted and interfering signals.

3.6.2 International frequency band planning

There are various forms of plan, drawn up internationally, for the arrangement of emissions within specific frequency bands. Typically, such plans cover frequency bands where radio propagation conditions lead to long interference distances and where system parameters are uniform and change only slowly. The other common application is for mobile systems, where the value of planning is mainly operational. The three basic forms of plan, namely radio channelling, frequency allotment and frequency assignment plans, are described briefly in Section 3.2.3.

Planning of this kind can make an important contribution to efficient spectrum utilisation. It provides a means of requiring radio users to implement a set of stringent spectrally efficient operational practices and technical standards in a specified frequency band if they are demonstrably necessary to enable that band to carry the traffic load which is required of it. Also frequency allotment and frequency assignment plans provide what may be an efficient method for optimising the geographical pattern of use of the channels in a frequency band. Plans also offer a basis for firm, mutual inter-governmental undertakings to respect the rights of foreign stations to use specified frequency assignments.

A frequency assignment planning conference may also be a helpful forum at which an administration with an immediate requirement for a new assignment in a crowded frequency band can negotiate adjustments to established assignments which will make way for the new station.

Frequency assignment or frequency allotment planning may also be used to reserve a share of a frequency band, associated in the case of satellite services with an arc of the geostationary satellite orbit, for the exclusive use of specified, and typically every, country. Such a plan is sometimes drawn up for a frequency band which is not yet in general use to avoid the possibility that a country, at some time in the future, should find it impossible to get access to the band for a new system, because the whole band has by that time become filled by the systems of other countries. Recent examples of such planning are the frequency assignment plans for satellite broadcasting that were agreed in 1977 and 1983 (see Section 3.4.6.2) and the spectrum/orbit allotment plan agreed in 1988 for the fixed-satellite service (see Section 3.4.2.6).

However, no rational and acceptable basis has ever been devised for sharing out a band between the nations. The actual requirements for which these reserved parcels of spectrum might ultimately be assigned are not known at the time of planning and many may never arise. And the systems, possibly few, that are established within the terms of such plans may be made more costly by the high standards of equipment performance which are demanded in order to provide for the possibility of access by a much larger number of systems, many of which are never implemented. Planning for this purpose meets a need that many administrations feel acutely, but it does nothing for economy in system costs or spectrum use for the planned service, and it may significantly impede the use of the same band for sharing services.

3.6.3 International regulations/recommendations

In certain radio services and in specified frequency bands, as indicated in Section 3.6.2, sets of spectrum-efficient operational practices and technical standards have been adopted as part of a plan. Exceptionally, in the fixed-satellite service using geostationary satellites, which is ill suited in precise planning, a similar though less comprehensive set of international practices and standards is being adopted in all bands as they come into full use, to raise the efficiency with which this service uses its transmission medium. See in particular the ITU Radio Regulations (ITU, 1990, Article 29) and various ITU-R recommendations (ITU, 1994). Some of the more significant of these measures are as follows.

There are regulations requiring frequency co-ordination between geostationary satellite networks, to ensure that the orbital separation between satellites can be minimised. The level of interference which is to be permitted, as the basis of co-ordination, has been raised over the years, allowing the angular separation between satellites to be further reduced so that more satellites can operate satisfactorily. There are regulations setting tolerances on satellite East West station keeping (ITU, 1990, Article 29), typically at 0.1^{o}. The accuracy of satellite antenna beam pointing is to be maintained within 10% of the half power beamwidth or 0.3^{o}, whichever is the greater.

However, as is apparent from Figures 3.4 to 3.6, interference between satellite networks arises largely from unwanted radiation of energy from antennas in the transmit mode (mainly satellite antenna radiation to locations outside the service area and earth station antenna radiation towards arcs of the geostationary satellite orbit away from where the wanted satellite is) and corresponding sensitivities to interference. In frequency co-ordination it may be possible to reduce such interference to an acceptable level by increasing the orbital separation between the satellites, but this option is of declining value as the number of satellites in use increases. In order to ensure that full use can be made of the medium, these entries of interference must be reduced by using earth station and satellite antennas with low sidelobe gain and by designing satellite networks to minimise the spectral power density of sidelobe radiation.

The ITU-R recommends limits on the spectral power density radiated off beam by transmitting earth stations. For example, at 6GHz the e.i.r.p. due to a wide band emission radiated at φ degrees off axis should not exceed $(35 - 25 \log \varphi)$ dBW in any sampling bandwidth of 4kHz when φ lies between 2.5^o and 48°. The corresponding limit for off axis angles greater than 48° is minus 7dBW per 4kHz. More stringent limits apply to antennas installed since 1988.

Also, design objectives have been set for earth station antenna sidelobe gain levels. For example, the ITU-R recommends for a new antenna, 5 metres in diameter, operating at 14GHz, that the maximum gain of at least 90% of the sidelobes should not exceed the values indicated in Figure 3.9 within the solid angle indicated in Figure 3.10. It is likely that corresponding targets will be set soon for satellite antenna beam overspill suppression.

3.6.4 National spectrum optimisation

Two kinds of radio facility, which are needed nowadays in great quantity in some countries, are fixed microwave links and vehicle borne and handportable radiotelephones. Indeed, ensuring that such systems use spectrum efficiently, so that the demand for frequency assignments can be met from finite bandwidth resources in parts of the spectrum where the radio propagation characteristics are suitable, may be the most difficult task that administrations face. However, cross frontier interference is not a major problem for these systems

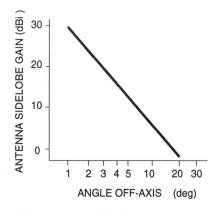

Figure 3.9   Recommended design objectives for the sidelobe gain of large earth station antennas installed after 1988. The maximum gain of 90% of the sidelobes within the solid angle shown in Figure 3.10 should not exceed $(29 - 25\log \varphi)$dBi where φ is the off-axis angle in degrees

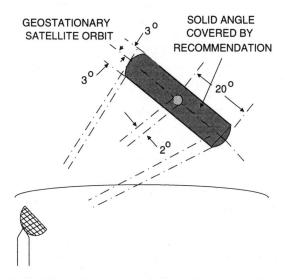

Figure 3.10   The solid angle to which the design objectives indicated in Figure 3.9 relate

and it has not been considered appropriate for the ITU to draw up international regulations or guidelines for the management of the spectrum that they use. Each administration, or group of administrations with a common policy in mobile radio, must find its own solutions.

A third major area of work for administrations is co-ordinating the frequencies assigned to earth stations (typically of the fixed satellite service) and terrestrial stations (typically of the fixed service) in shared frequency bands.

These three problem areas are touched on earlier but they are considered in rather more detail below.

3.6.4.1 *Fixed microwave links*

The main elements of efficient planning for fixed links are:

1. Definition of radio channelling plans (such as, for example, Figure 3.1) for the available frequency band.
2. Publication of technical specifications for type approval of transmitting and receiving equipment, compatible with the radio channel plans. The specification would set standards of such factors as antenna sidelobe response, antenna polarisation performance and rejection of adjacent channel interference. These standards would be stringent if circumstances made high efficiency of spectrum utilisation necessary.
3. Determination of circuit noise and interference objectives. The target level for noise plus interference in channels should not be made unnecessarily good, and the ratio of interference to noise might favour the former if efficient spectrum utilisation was important.
4. Channels would then be assigned to links as the need arises, the transmitter power being determined in each case to ensure that the noise plus interference performance realised did not too greatly exceed the objective.

The key operation in this procedure is the assigning of channels to links. When a substantial number of assignments has been made, each frequency may be re-used several or many times over, with geographical separation. The separation may be quite small if good use can be made of the directional properties of the antennas. Computer software which is available does this task well, but digital maps giving very precise topographical data are necessary if the highest efficiency in the use of spectrum is to be attained.

3.6.4.2 *Mobile radiotelephones*

For mobile radiotelephone facilities, different problems arise in managing spectrum. Here again a radio channelling plan is required. Channel frequencies will be re-used with geographical separation between service areas. The basic equipment type approval specification will emphasise:

1. Excellent suppression of spurious emissions from mobile transmitters and receivers.
2. Low adjacent channel emission in transmitters and low adjacent channel response in receivers.
3. Optimised channel bandwidth requirement, typically at present in the range 5kHz to 25kHz, and corresponding carrier frequency stability.
4. Freedom from receiver blocking by powerful signals off the tuning point.

However, traffic in the land mobile service typically consists of brief conversations between pairs of users, interrupted by periods of silence that may be long. The key to efficient spectrum utilisation lies in ensuring that a channel is not idle during the periods of silence. The nature of the user's business will determine what system architecture will provide efficient spectrum loading at acceptable cost.

For many radiotelephone users, access to the public switched telephone network is the basic requirement. There are cellular systems to meet this need in many countries and especially in the more densely populated areas. However, for other users much more limited access would be sufficient for their business, and it may be possible to obtain this from private networks at less than the cost of cellular systems.

A busy vehicle fleet operator, passing perhaps hundreds of short messages per hour to taxicabs, police cars etc. operating within an area some tens of kilometres in radius, may well make efficient use of spectrum as the sole assignee of a channel, using a simple transmitter/receiver network.

An operator serving a small fleet of vehicles could seldom justify the exclusive use of a channel and a private user needing access from one vehicle would never do so. The level of use of the channel would usually be raised by assigning the same channel in the same area to several users. Simple control systems are available to stop one assignee from starting to transmit at a time when another assignee is using the channel. This provides a low cost system, acceptable in spectrum utilisation efficiency in locations where the demand for spectrum is not high, although users find shared channels irksome.

In locations where demand is high, systems which offer access for each user to whichever of many channels is free at the moment when a call is to start can combine convenience to the user with very efficient spectrum utilisation. Multi-channel trunked systems might be suitable if the vehicles remain for most of the time within range of a single base station. Private cellular systems are necessary if the mobile station roams widely.

Given sufficient need for efficient spectrum utilisation, the spectrum manager must direct the users' choice of system towards appropriate system architecture.

3.6.4.3 *Frequency co-ordination around earth stations*

In shared frequency bands allocated to both the fixed and fixed-satellite services, a frequency co-ordination procedure is used to limit interference between earth stations and nearby terrestrial stations. The procedure is outlined in Section 3.4.2.5. above. It is labour intensive but its application is quite feasible given the assumptions for which it was developed, namely that there are relatively few earth stations. However substantial numbers of new earth stations are now being installed, mostly Very Small Aperture Terminals (VSATs). The use of frequency co-ordination to manage spectrum in bands shared by large numbers of stations of both space and terrestrial services is likely to lead to inefficient use of spectrum for one or both services and spectrum management costs may become disproportionate to the value of the telecommunication facilities made available.

Differences in the configuration of microwave terrestrial and satellite networks are a key factor here. If spectrum that is shared between these two services is to be used efficiently by both, the receivers of one service must be protected from the transmitters of the other by geographical separation or by obstacles in the direct propagation path between them. Now it is necessary for the antennas of most terrestrial stations to be mounted high on the skyline, on towers, the roofs of high buildings and hill tops; only in this way can line of sight propagation for terrestrial links of substantial length be obtained. Also, many terrestrial stations are located in city centres. There is no corresponding need for earth station antennas to be mounted high, and interference is reduced and co-ordination simplified if they are low. In considering how this basic factor affects practical spectrum management, three typical situations can be identified.

The first situation arises where the earth stations are used to provide high capacity network links, typically for international or trunk telecommunication connections or for television programme feeds, upwards to broadcasting satellites or downwards to terrestrial broadcasting transmitters or cabled networks. Such earth stations can be sited with due care, in locations which are remote from terrestrial stations or screened from terrestrial station antennas, typically by hills or massive buildings. Conventional frequency co-ordination is appropriate in such a situation and is in general use.

The second situation has evolved from the use of satellites operating in frequency bands allocated to the fixed-satellite service for distributing television programmes to terrestrial broadcasting stations and cabled networks. Large numbers of small receiving antennas are now being used domestically in some countries to intercept these signals for home entertainment. In many residential localities interference from terrestrial transmitters may not be unacceptably high, but where interference is strong, there is no practical way in which frequency co-ordination can be used to reduce it.

The third situation concerns VSATs. Some of these earth stations are installed at remote locations, where interference to or from terrestrial stations may not be serious and conventional frequency co-ordination may be quite feasible, although the cost may be disproportionately high, relative to the value of the facilities provided by the VSAT station. However, installation of VSATs at city centre locations is more usual and rooftop antenna mounting would often be preferred. Spectrally efficient frequency co-ordination between these city centre VSATs and terrestrial systems operating from city centre towers is often not feasible.

The current indications are that the total bandwidth that VSAT networks will occupy in the foreseeable future may not be very great; a few hundreds of megahertz may well be ample. If so, the best arrangement would be for VSAT networks to be assigned frequencies in fixed satellite allocations which are not shared in the earth station locality with terrestrial services. There are few unshared frequency allocations for the fixed satellite service in the international table of frequency allocations. The best prospects are as follows:

1. Up links; 14.0 to 14.5GHz (Region 2); 14.0 to 14.25GHz (Regions 1 and 3); 29.5 to 30.0GHz (all Regions).
2. Down links: 12.1 to 12.2GHz (Region 2); 12.5 to 12.75GHz (Region 1); 19.7 to 20.2GHz (all Regions).

Even these narrow bands are allocated to terrestrial services in very large numbers of countries by footnotes to the international table. However, one highly significant footnote (ITU, 1990, para 837) to the table reduces the terrestrial allocations in the band 11.7GHz to 12.1GHz to secondary status in Canada, Mexico and the U.S.A., effectively extending to 11.7GHz to 12.2GHz in those countries the band where earth station assignments do not have to be co-ordinated. Other countries, where footnotes to the international table provide a terrestrial allocation in the bands listed above, may decide not to activate those allocations nationally if substantial use of VSAT networks is foreseen.

3.6.5 Market forces in spectrum management

So far it has been assumed that the criterion for effectiveness in spectrum utilisation is enabling the largest quantity of radio services of all useful kinds to operate effectively in a large but not limitless spectrum. The means that are used for achieving this objective have been regulatory and technical. However, it has been argued that other objectives are to be preferred.

Economists point out that radio spectrum is exceptional among factors of substantial economic importance in that it is available at virtually no cost. Levin (1971), for example, argues that in consequence there is no satisfactory mechanism for ensuring that spectrum is distributed between users in an economically optimum way. The application of market forces should lead to spectrum being used, not necessarily to provide the greatest quantity of radio services, but for radio services of the greatest aggregate market value. He recognises that the establishment of a completely free market in spectrum presents special problems and is probably impracticable but he concludes that some substantial application of market forces would be beneficial.

The New Zealand government introduced a new spectrum management regime for some frequency bands in 1989 which involved elements of a market system. However, the use of market forces in this way seems indeed to present major problems. Any system which takes authority over radio systems which can cause international interference out of the hands of governments raises problems, solutions to which are not evident. Also, loss of the benefits of internationally standardised frequency allocation in such areas as mobile communication and radionavigation would cause widespread difficulties and dangers. Furthermore, it is not obvious that facilities that only radio can provide can properly be denied to society because their market value is small.

However, there are situations where economic pressures of some kind would seem to be justifiable. In particular, the decision to use a microwave link between fixed points, instead of a cable connection, will often be influenced by the fact that cabling may be costly. If there is plenty of spectrum, the use of radio is a welcome benefit to the user.

Where spectrum is in great demand, however, a case could be made for adding a leasing charge to the licence fee to encourage frugal use of bandwidth and to ensure that cables are used where economically feasible, leaving radio for the user for whom cable is a particularly costly option.

3.7 References

HMSO (1985) *United Kingdom Table of Radio Frequency Allocations*.

ITU (1983) *Appendix 27, Radio Regulations*. The International Telecommunications Union, Geneva.

ITU (1990) *Radio Regulations*, The International Telecommunication Union, Geneva.

ITU (1992a) *Final Acts of the WARC for dealing with frequency allocations in certain parts of the spectrum*, ITU, Geneva.

ITU (1992b) *Final Acts of the Additional Plenipotentiary Conference (APP-92)*, ITU, Geneva.

ITU (1994) *Recommendations of the ITU Recommendations Assembly (ITU-R) in force in 1994*. ITU, Geneva.

Levin, H.J. (1971) *The invisible resource, use and regulation of the radio spectrum*. The Johns Hopkins Press.

4. Antennas

4.1 Types of antennas

Antennas form the link between the guided parts and the free-space parts of a communication system. The purpose of a transmitting antenna is to efficiently transform the currents in a circuit or waveguide into radiated radio or microwave energy. The purpose of a receiving antenna is to efficiently accept the received radiated energy and convert it to guided form for detection and processing by a receiver. The design and construction of an antenna usually involves compromises between the desired electromagnetic performance and the mechanical size, mass and environmental characteristics.

Antennas for communication systems fall into two broad categories depending on the degree to which the radiation is confined. Microwave radio relay and satellite communications use pencil beam antennas, where the radiation is confined to one narrow beam of energy, Figure 4.1. Mobile communications are more likely to require antennas with omni-directional patterns in the horizontal plane and toroidal patterns in the vertical plane, Figure 4.2.

Pencil beam antennas usually consist of one or more large to medium reflectors which collimate the signals from a feed horn at the focus of the reflector. Both reflector and feed horn fall within the generic class of aperture antennas because they consist of an aperture

Figure 4.1   Pencil beam radiation pattern

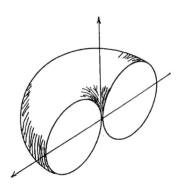

Figure 4.2   Toroidal radiation pattern

which radiates into space. The design problem is to determine first the aperture fields which will yield the specified radiation characteristics and secondly to design the reflectors and horns to produce the aperture fields. Aperture antennas can be designed to meet very stringent specifications. Omni-directional antennas consist of elements which are small in wavelengths, such as dipoles and monopoles. The radiation characteristics are influenced by the presence of surrounding objects. Non-electromagnetic factors such as the size are often as important in the design as the radiation performance. For this reason the design of omni-directional antennas is partly an empirical process in which expertise and previous experience play an equal part with theoretical knowledge.

In between the large aperture antennas and the small element antennas lie array antennas which consist of two or more elements. The radiation from an array antenna is determined principally by the physical spacing and electrical signals driving the elements rather than the radiation characteristics of the elements themselves.

The detailed theory of antennas can be found in Stutzman (1981); Elliot (1981); Balanis (1982); Silver (1984); Collin (1969). Design information and descriptions of particular types can be found in a number of handbooks, such as, Rudge (1986); Johnson (1984); Lo (1988); Milligan (1985).

4.1.1 Antennas used in communications

Table 4.1 lists the principal types of antennas which are used in communications. Under each category of communication system, the specific type of antenna which is usually used is given. The last column gives the generic type of either an aperture antenna, an array antenna or a small element antenna. The generic type describes the general radiation characteristics which can be obtained from the antenna and is useful because it makes the explanation of the performance easier.

The following sections describe first the generic antenna characteristics, then the specific antenna types and then a brief discussion of the practical implementation of antennas in communication systems.

4.2 Basic properties

The principle of reciprocity is one of the most important properties of an antenna. It means that the properties of an antenna when acting as a transmitter are identical to the properties of the same antenna when acting as a receiver. For this to apply, the medium in between the two antennas must be linear, passive and isotropic, which is always the case for communication systems.

The directional selectivity of an antenna is represented by the radiation pattern. It is a plot of the relative strength of the radiated field as a function of the angle. A pattern taken along the principal direction of the electric field is called an E-plane cut, the orthogonal plane is called an H-plane cut. The most common plot is the rectangular decibel plot, Figure 4.3 which can have scales of relative power and angle chosen to suit the antenna being characterised. Other types of plots such as polar plots (used for small antennas and two dimensional), contour plots (or three dimensional), and isometric plots are also used. A radiation pattern is characterised by the main beam and sidelobes. The quality is specified by the beamwidth between the −3dB points on the main beam and the sidelobe level.

Communication antennas radiate in either linear polarisation or circular polarisation. In modern communications cross-polarisation is important. This is the difference between the two principal plane

Table 4.1 Antennas used in communication systems

Use	Specific type	Generic type
Microwave line of sight radio	Prime focus reflector with small feed	Apertures
Earth Stations (Large)	Dual reflector with corrugated horn feed	Apertures
Earth Stations (Medium)	Offset reflector with corrugated or dual mode horn feed	Apertures
Direct Broadcast Satellite Receiving Antennas	Prime focus symmetric or offset reflector	Apertures
	Flat plate antennas	Arrays
Satellite Antennas (Spot beams)	Offset reflector with single feed	Apertures
Satellite Antennas (Multiple beams)	Offset reflector with array feed	Aperture and Arrays
Satellite Antennas (Shaped beams)	Shaped reflectors	Apertures
	Offset reflector with array feed	Apertures and Arrays
VHF/UHF Communications	Yagis Dipole arrays Slots	Elements and Arrays
Mobile Communications (Base stations)	Dipole arrays	Elements and Arrays
Mobile Communications (Mobile)	Monopoles Microstrip	Elements
HF Communications	Dipoles Monopoles	Elements

patterns and is specified relative to a reference polarisation, called the co-polar pattern. There are three definitions of the cross-polarisation. The one in normal use with reflector antennas and feed systems is

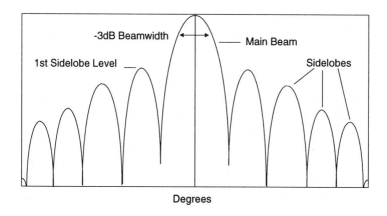

Figure 4.3   Typical rectangular radiation plot

Ludwig's third definition (Ludwig, 1973) which assumes that the reference polarisation is that due to a Huygens source. It most closely corresponds to what is measured with a conventional antenna test range.

The power gain in a specified direction is defined by the ratio of the power radiated per unit solid angle in direction θ,φ to the total power accepted from the source, as in Equation 4.1.

$$G(\theta,\varphi) =$$

$$4\pi\frac{Power\ radiated\ per\ solid\ angle\ in\ direction\ \theta,\varphi}{Total\ power\ accepted\ source} \qquad (4.1)$$

This is an inherent property of an antenna and includes dissipative losses in the antenna. The dissipative losses cannot easily be predicted so a related parameter, the directivity, is used in calculations. The definition of the directivity is similar to that of the gain except that the denominator is replaced by the total power radiated. The terms gain and directivity are often used interchangeably in the literature. Normally only the peak gain along the boresight direction is specified. If the direction of the gain is not specified, the peak value is assumed. The value is normally quoted in dBs. The definitions given above are in effect specifying the gain relative to a loss-less isotropic source. This is sometimes stated explicitly by using the symbol dBi.

The efficiency of an aperture antenna is given by the ratio of the effective area of an aperture divided by the physical area. Normal aperture antennas have efficiencies in the range 50-80%.

As far as circuit designers are concerned the antenna is an imped-ance. Maximum power transfer will occur when the antenna is matched to the transmission line. The impedance consists of the self impedance and the mutual impedance. The mutual impedance ac-counts for the influence of nearby objects and of mutual coupling to other antennas.

The self impedance consists of the radiation resistance, the loss resistance and the self reactance. Loss resistance is the ohmic losses in the antenna structure. Radiation resistance measures the power absorbed by the antenna from the incoming plane waves. It is one of the most significant parameters for small antennas where the problem is often to match very dissimilar impedances.

A receiving antenna is both a spatially selective filter (measured by the radiation pattern) and a frequency selective filter. The bandwidth measures the frequency range over which the antenna operates. The upper and lower frequencies can be specified in terms of a number of possible parameters: gain, polarisation, beamwidth and impedance.

A communication link consists of a transmitting antenna and a receiving antenna. If the transmitter radiates P_t watts, then the re-ceived power, P_r at a distance r is given by Equation 4.2 where G_t and G_r are the transmitter and receiver antenna gains respectively.

$$P_r = P_t \frac{G_t G_r \lambda^2}{(4\pi r)^2} \tag{4.2}$$

This formula is known as the Friis transmission equation. It as-sumes that the antennas are impedance and polarisation matched. If this is not the case then extra factors must be multiplied to the equation to account for the mismatches.

4.3 Generic antenna types

4.3.1 Radiation from apertures

The radiation from apertures illustrates most of the significant properties of pencil beam antennas. The radiation characteristics can be
determined by simple mathematical relationships. If the electric
fields across an aperture, Figure 4.4, is $E_a(x,y)$ then the radiated fields
$E_p(\theta,\varphi)$ is given by Equation 4.3, where $f(\theta, \varphi)$ is given by Equation 4.4. (Olver, 1986; Milligan, 1985).

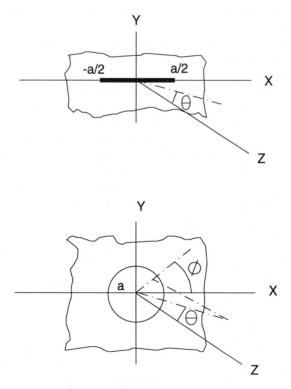

Figure 4.4   Radiation aperture in a ground plane

$$E_p(\theta,\varphi) = \cos^2\frac{\theta}{2}\left(1 - \tan^2\frac{\theta}{2}\cos 2\varphi\right) f(\theta,\varphi) \tag{4.3}$$

$$f(\theta,\varphi) = \int\limits_{-\infty}^{\infty}\int\limits_{\infty}^{\infty} E_a(x,y)\, e^{jk(x\sin\theta\cos\varphi + y\sin\theta\sin\varphi)} dx\, dy \tag{4.4}$$

For high or medium gain antennas the pencil beam radiation is largely focused to a small range of angles around $\theta = 0$. In this case it can be seen from Equation 4.3 that the distant radiated fields, and the aperture fields are the Fourier transformation of each other.

Fourier transforms have been widely studied and their properties can be used to understand the radiation characteristics of aperture antennas. Simple aperture distributions have analytic Fourier transforms, whilst more complex distributions can be solved numerically on a computer.

The simplest aperture is a one dimensional line source distribution of length $\pm\dfrac{a}{2}$. This serves to illustrate many of the features of aperture antennas. If the field in the aperture is constant, the radiated field is given from Equation 4.3 as in Equations 4.5 and 4.6.

$$E_p = \frac{\sin(\pi u)}{\pi u} \tag{4.5}$$

$$u = \frac{a}{\lambda}\sin\theta \tag{4.6}$$

This distribution occurs widely in antenna theory. It is plotted in Figure 4.5. The beamwidth is inversely proportional to the aperture width and is $\dfrac{0.88\lambda}{a}$. The first sidelobe level is at −13.2dB which is a disadvantage of a uniform aperture distribution. The level can be reduced considerably by a tapered aperture distribution where the field is greatest at the centre of the aperture and tapers to a lower level at the edge of the aperture. For example if Equation 4.7 holds, then the first sidelobe level is at −23dB.

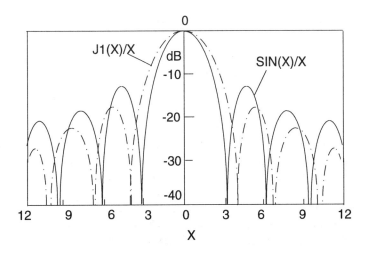

Figure 4.5   Dipole and radiation pattern

$$E_a(x) = \cos\left(\frac{\pi x}{2a} \right) \tag{4.7}$$

The energy which was in the sidelobes moves to the main beam with the result that the beamwidth broadens to $\frac{1.2\lambda}{a}$. In practice almost all antennas have natural tapers across the aperture which result from boundary conditions and waveguide modes. Rectangular apertures are formed from two line source distributions in orthogonal planes.

Circular apertures form the largest single class of aperture antennas. The parabolic reflector is widely used in communications and is often fed by a conical horn. Both the reflector and the horn are circular apertures. For an aperture distribution which is independent of azimuthal angle the simplest case is uniform illumination which gives a radiated field as in Equation 4.8, where $J_1(x)$ is a Bessel function of zero order.

$$E_p = \frac{2\,J_1\,(\,\pi u\,)}{\pi u} \qquad\qquad (4.8)$$

This can be compared to $\frac{\sin(x)}{x}$ and is also plotted in Figure 4.5. The first sidelobe level is at −4.6dB. Table 4.2 lists a number of circular aperture distributions and corresponding radiation pattern properties. The pedestal distribution is representative of many reflector antennas which have an edge tapers of about −10dB corresponding to $E_a(a) = 0.316$. The Gaussian distribution is also important

Table 4.2 Radiation characteristics of circular apertures

Electric field aperture distribution	3dB beamwidth	Level of first sidelobe
Uniform	$1.02\,\dfrac{\lambda}{D}$	−17.6dB
Taper to zero at edge $1-\left(\dfrac{2r}{D}\right)^2$	$1.27\,\dfrac{\lambda}{D}$	−24.6dB
Taper on a pedestal $0.5+\left[\,1-\left(2\dfrac{r}{D}\right)^2\,\right]^2$	$1.16\,\dfrac{\lambda}{D}$	−26.5dB
Gaussian $\exp\left[\,-p\left(\dfrac{2r}{D}\right)^2\,\right]$	$1.33\,\dfrac{\lambda}{D}$	−40dB (p=3)

because high performance feed horns ideally have Gaussian aperture distributions. The Fourier transform of a Gaussian taper which decreases to zero at the edge of the aperture gives a Gaussian radiation pattern which has no sidelobes.

4.3.2 Radiation from small antennas

Small antennas are needed for mobile communications operating at frequencies from HF to the low microwave region. Most of these are derivatives of the simple dipole, Figure 4.6, which is an electric current element which radiates from the currents flowing along a small metal rod. The radiation pattern is always very broad with energy radiating in all directions. An important design parameter is the impedance of the dipole which can vary considerably depending on the exact size and shape of the rod. This means that the impedance

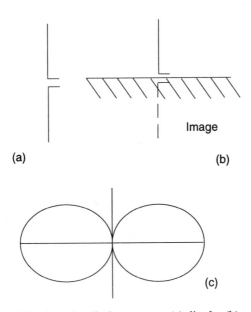

Figure 4.6   Dipole and radiation pattern: (a) dipole; (b) monopole; (c) polar pattern of dipole

matching between the antenna and the transmitting or receiving circuit becomes a major design constraint.

The radiation fields from a dipole are obtained by integrating the radiation from an infinitesimally small current element over the length of the dipole. This depends on knowing the current distribution which is a function not only of the length but also of the shape and thickness of the rod. Many studies have been addressed towards obtaining accurate results (King, 1956; King, 1968). For most cases this has to be done by numerical integration. A simple case is a short dipole with a length $a \ll \dfrac{1}{10\lambda}$ when the current distribution may be assumed to be triangular. This results in radiated fields of the form given in Equation 4.9.

$$E = j \, 30 \, \pi \, I \, \frac{a}{\lambda} \, \sin \theta \, \frac{e^{-j k r}}{r} \tag{4.9}$$

The electric field is plotted in polar form in Figure 4.6. The radiation resistance is calculated by evaluating the radiated power and using $P = I^2 R$ to give Equation 4.10.

$$R = 20\pi^2 \left(\frac{a}{\lambda} \right)^2 \tag{4.10}$$

A dipole of length $a = \dfrac{\lambda}{10}$ has a radiation resistance of 2.0 ohms. This is low by comparison with standard transmission lines and indicates the problem of matching to the transmission line.

The half wave dipole is widely used. Assuming a sinusoidal current distribution the far fields are given by Equation 4.11.

$$E = j \, 60 \, I \, \frac{\cos [\, \pi / 2 \cos \theta \,]}{\sin \theta} \frac{e^{-j k r}}{r} \tag{4.11}$$

This gives a slightly narrower pattern than that of the short dipole and has a half beamwidth of 78 degrees. The radiation resistance must be evaluated numerically. For an infinitely thin dipole it has a value of 73 + j 42.5 ohms. For finite thickness the imaginary part can become zero in which case the dipole is easily matched to a coaxial cable of impedance 75ohms. The half wave dipole has a gain of 2.15dB.

A monopole is a dipole divided in half at its centre feed point and fed against a ground plane, Figure 4.6. The ground plane acts as a mirror and consequently the image of the monopole appears below the ground. Since the fields extend over a hemisphere the power radiated and the radiation resistance is half that of the equivalent dipole with the same current. The gain of a monopole is twice that of a dipole. The radiation pattern above the ground plane is the same as that of the dipole.

4.3.3 Radiation from arrays

Array antennas consist of a number of discrete elements which are usually small in size. Typical elements are horns, dipoles, and microstrip patches. The discrete sources radiate individually but the pattern of the array is largely determined by the relative amplitude and phase of the excitation currents on each element and the geometric spacing apart of the elements. The total radiation pattern is the multiplication of the pattern of an individual element and the pattern of the array assuming point sources, called the array factor. Array theory is largely concerned with synthesising an array factor to form a specified pattern. In communications most arrays are planar arrays with the elements being spaced over a plane, but the principles can be understood by considering an array of two elements with equal amplitudes, Figure 4.7(a). This has an array factor given by Equation 4.12, where Ψ is given by Equation 4.13.

$$E = E_1 + E_2 \, e^{j\,\Psi} \tag{4.12}$$

$$\Psi = \delta + k \, d \cos \theta \tag{4.13}$$

The pattern for small spacings will be almost omnidirectional and as the spacing is increased the pattern develops a maxima perpendicu-

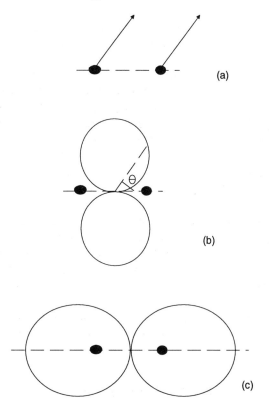

Figure 4.7   Two element array and radiation pattern: (a) two elements with equal amplitudes; (b) half wavelength spacing; (c) 180° phase difference

lar to the axis of the array. At a spacing of half a wavelength, a null appears along the array axis, Figure 4.7(b). This is called a broadside array. If a phase difference of 180 degrees exists between the two elements then the pattern shown in Figure 4.7(c) results. Now the main beam is along the direction of the array and the array is called the end-fire array. This illustrates one of the prime advantages of the array, namely by changing the electrical phase it is possible to make the peak beam direction occur in any angular direction. Increasing the spacing above half a wavelength results in the appearance of additional radiation lobes which are generally undesirable. Consequently the ideal arrays spacing is half wavelength, though if waveguides or horns are used this is not usually possible because the basic element is greater than half a wavelength in size. Changing the relative amplitudes, phases and spacings can produce a wide variety of patterns so that it is possible to synthesise almost any specified radiation pattern. The array factor for an N element linear array of equal amplitude is given by Equation 4.14.

$$E = N \frac{\sin (N \Psi / 2)}{\sin (\Psi / 2)} \qquad (4.14)$$

This is similar to the pattern of a line source aperture, Equation 4.5, and it is possible to synthesise an aperture with a planar array. There is a significant benefit to this approach. The aperture fields are determined by the waveguide horn fields which are constrained by boundary conditions and are usually monotonic functions. This constraint does not exist with the array so that a much larger range of radiation patterns can be produced. Optimum patterns with most of the energy radiated into the main beam and very low sidelobes can be designed.

4.4 Specific antenna types

4.4.1 Prime focus symmetric reflector antennas

4.4.1.1 *Parabolic reflectors*

The axi-symmetric parabolic reflector with a feed at the focus of the paraboloid is the simplest type of reflector antenna. The geometry is shown in Figure 4.8. The paraboloid has the property that energy from the feed at F goes to the point P on the surface where it is reflected parallel to the axis to arrive at a point A on the imaginary aperture plane.

The equation describing the surface is given by Equation 4.15, where F is the focal length.

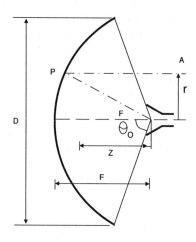

Figure 4.8   Geometry of reflector antenna

$$r^2 = 4 F (F - z)$$ (4.15)

At the edge of the reflector, of diameter D, Equation 4.16 applies.

$$\frac{F}{D} = \frac{1}{4} \cot \left(\frac{\theta_0}{2} \right)$$ (4.16)

The depth of the paraboloid is usually specified by its F/D ratio. Common sizes are between F/D = 0.25 (θ_0 = 90 degrees) to F/D = 0.4.

4.4.1.2 *Aperture fields and radiation patterns*

Ray optics indicates that the path length from F to P to A is equal to twice the focal length. Hence the phase across the aperture is constant. The amplitude across the aperture plane will peak at the centre and taper towards the edge for two reasons. Firstly because the feed will have a tapered radiation pattern and secondly because the action of a parabola in transforming a spherical wave from the feed into a plane wave across the aperture introduces a path loss which is a function of angle θ. The aperture electric field is then given by Equation 4.17, where $F(\theta,\varphi)$ is the pattern of the feed.

$$E_a(\theta,\varphi) = F(\theta,\varphi) \cos^2\left(\frac{\theta}{2}\right) \tag{4.17}$$

Feeds suitable for reflector antennas are discussed in a later section, but it is often convenient in initial design to take the feed pattern as being given by Equation 4.18.

$$F(\theta,\varphi) = \cos^q(\theta) \tag{4.18}$$

Experience has shown that good quality feeds approximate well to this function.

The radiation patterns can be predicted from the aperture fields by using the Fourier transform relations described in Section 4.3.1. This works well for large reflectors but for detailed design of small to medium reflectors it is necessary to take account of the precise form of the currents on the reflector surface and the diffraction that occurs at the edges of the reflector surface. The former can be accomplished with Physical Optics theory (Rusch, 1970; Rusch, 1986) and is good for predicting the main beam and near-in sidelobes. The diffracted fields influence the far-out sidelobes and can be predicted using the Geometrical Theory of Diffraction (GTD) (James, 1986).

4.4.1.3 *Gain of reflector antennas*

The gain of a reflector antenna can be calculated from Equation 4.19, where η is the efficiency of the reflector.

$$G = \eta\left(\frac{\pi D}{\lambda}\right)^2 \tag{4.19}$$

The total efficiency is the product of six factors:

1. The illumination efficiency is the gain loss due to the non-uniform aperture illumination.

2. The spillover efficiency is the gain loss caused by energy from the feed which radiates outside the solid angle subtended by θ_0 called the spillover. It is the fraction of the power which is intercepted by the reflector. As the aperture edge taper increases, the spillover decreases and the spillover efficiency increases, whilst the illumination efficiency decreases. There is an optimum combination which corresponds to an edge illumination of about $-10\mathrm{dB}$.

3. The phase error efficiency is a measure of the deviation of the feed face front away from spherical and is usually nearly 100%.

4. The cross-polarisation efficiency is a measure of the loss of energy in the orthogonal component of the polarisation vector. For a symmetric reflector no cross-polarisation is introduced by the reflector so the efficiency is determined by the feed characteristics. For good feeds this factor is also nearly 100%.

5. The blockage efficiency is a measure of the portion of the aperture which is blocked by the feed and the feed supports. The fields blocked by the feed do not contribute to the radiation so it is desirable to keep the proportion of the area blocked to less than 10% of the total area of the aperture because otherwise the sidelobe structure becomes distorted. The feed support blocking is more complicated because it depends on the shape and orientation of the supports (Lamb, 1986). It is electrically desirable to keep the cross-section of the supports small which means that a compromise with the mechanical constraints is needed.

6. The surface error efficiency is a measure of the deviations of the aperture wavefront from a plane wave due to surface distortions on the parabolic surface. Assuming that the errors are small and randomly distributed with a root mean square (r.m.s.) surface error, the efficiency is given by Equation 4.20. This is a function of frequency and falls-off rapidly above a certain value which means that the upper frequency for which a reflector can be used is always given by the surface errors.

The effect on the radiation pattern of random surface errors is to fill in the nulls and to scatter energy in all directions so that the far out sidelobes are uniformly raised.

$$\eta_S = \exp\left[-\left(\frac{4\pi\varepsilon}{\lambda} \right)^2 \right]$$

(4.20)

4.4.2 Dual symmetric reflector antennas

The performance of a large reflector antenna can be improved and the design made more flexible by inserting a sub-reflector into the system, Figure 4.9. There are two versions, the Cassegrain, where the sub-reflector is a convex hyperboloid of revolution placed on the

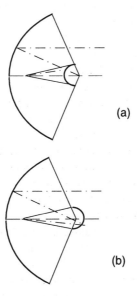

Figure 4.9 Dual reflector antennas: (a) Cassegrain; (b) Gregorian

inside of the parabola focus, and a Gregorian where a concave elliptical sub-reflector is placed on the outside of the parabola focus. In symmetric reflectors the Cassegrain is more common because it is more compact, but the electrical performance is similar for both systems.

The advantages of the dual reflectors are:

1. The feed is in a more convenient location.
2. Higher performance feeds can be used because the subtended angle is such that wide aperture diameter feeds are needed.
3. Spillover past the sub-reflector is directed at the sky which reduces the noise temperature.
4. The depth of focus and field of view are larger.

The study of the radiation characteristics and the efficiency of dual reflectors is similar to that for the prime focus reflector. Analysis of the radiation patterns depends partly on the size of the sub-reflector. If it is small then physical optics or GTD must be used on the sub-reflector. The main reflector is usually large so geometric optics is adequate.

The limiting factor to obtaining high efficiency in a standard parabola is the amplitude taper across the aperture due to the feed pattern and the space loss in the parabola (i.e. the illumination efficiency). By shaping the surfaces of a dual reflector antenna it is possible to increase the efficiency and produce a more uniform illumination across the aperture.

A well known method to produce a high efficiency Cassegrain symmetric reflector antenna is due to Galindo and Williams (Galindo, 1964; Williams, 1965). It is a geometric optics technique in which the shape of the sub-reflector is altered to redistribute the energy more uniformly over the aperture. Then the shape of the main reflector is modified to refocus the energy and create a uniform phase across the aperture. After this process the reflector surfaces are no longer parabolic and hyperbolic. The method works well for large reflectors. For small or medium size reflectors geometric optics is not adequate and physical optics including diffraction must be used at least on the sub-reflector.

4.4.3 Offset reflectors

In recent years the growth in communication systems has led to a tightening in the radiation pattern specifications and the consequent need to produce reflectors with low far-out sidelobes. Symmetric reflectors cannot be made to have low sidelobes because of the inherent limitations caused by scattering from the feed and feed supports. This blockage loss can be entirely eliminated with the offset reflector, Figure 4.10, which consists of a portion of a parabola chosen so that the feed is outside the area subtended by the aperture of the reflector. The projected aperture is circular, though the edge of the reflector will be elliptical. The removal of the blockage loss also means that smaller reflector antennas can be made efficient which has led to their widespread use as DBS receiving antennas.

In addition to the unblocked aperture, the offset reflector has other advantages, (Rudge, 1986, page 185; Rahmat-Samii, 1986). The reaction of the reflector upon the primary feed can be reduced to a very low order so that the feed VSWR is essentially independent of the reflector. Compared to a symmetric paraboloid, the offset con-

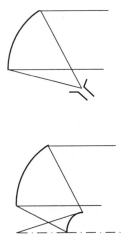

Figure 4.10   Offset reflector antennas

figuration makes a larger F/D ratio possible which in turn enables a higher performance feed to be used. The removal of the feed from the aperture gives greater flexibility to use an array of feeds to produce multiple beams or shaped beams.

The offset reflector antenna also has some disadvantages. It is much more difficult to analyse and design due to the offset geometry and it is only with the advent of powerful computers that this has become feasible. The lack of symmetry in the reflector means that when a linearly polarised feed is used, a cross-polarised component is generated by the reflector surface. When circular polarisation is used, a cross-polarised component does not occur but the offset surface causes the beam to be 'squinted' from the electrical boresight. Lastly the construction of the offset reflector is more difficult. However if the reflectors are made by fibreglass moulding this is not really significant. Also the structural shape can be put to good use because it is convenient for deployable configurations on satellites or transportable earth stations.

4.4.4 Horn feeds for reflector antennas

A reflector antenna consists of the reflector plus the horn feed at the geometric focus of the reflector. Thus the correct choice and design of the feed is an important part of the design of the total reflector antenna. High performance feeds are necessary to achieve high performance antennas. The diameter of the feed in wavelengths will be determined by the angle subtended by the reflector at the feed. A prime focus reflector with an F/D between 0.25 and 0.5 will have a subtended half angle of between 90 degrees and 53 degrees. Application of the general rule that beamwidth is approximately equal to the inverse of the normalised aperture diameter shows that this means a feed with an aperture diameter of between about one and three wavelengths. Dual reflectors (Cassegrain or Gregorian) and offset reflectors have subtended angles between 30 degrees and 7 degrees, leading to feed diameters of between three and ten wavelengths.

Of particular interest in horn feed design is the polarisation performance and the quality of a feed is usually expressed by the level of the peak cross-polarisation. The radiation characteristics of horns are

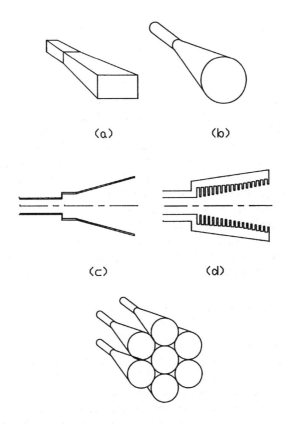

Figure 4.11 Types of feed horns for reflector antennas:
(a) rectangular or square; (b) small conical; (c) small conical with
chokes; (d) conical corrugated; (e) array feed

predicted by a two part process. Firstly the fields in the aperture are
computed from a knowledge of the guided wave behaviour inside the
horn. Secondly the aperture fields are used to compute the radiated
fields. The Fourier transform method has been found to work very
well for the case of horns. The main types will now be briefly
described. For more details see Love (1976 and Love (1986).

4.4.4.1 *Rectangular or square horns*

These are the simplest type of horn, Figure 4.11(a) but they are rarely used as feeds for reflectors because they have very high cross-polarisation unless the aperture size is large.

4.4.4.2 *Small conical horns*

These can have reasonably good cross polarisation performance, Figure 4.11(b). They are widely used as prime focus feeds in small symmetric and offset reflectors. The basic design will have an aperture diameter of about one wavelength and is essentially an open-ended circular waveguide propagating a TE_{11} mode. The radiation pattern can be improved by adding one or more rings or chokes around the aperture, Figure 4.11(c). These have the effect of changing the distribution of current on the flange and creating a more symmetric radiation pattern. The theoretical design of the open-ended waveguide is straightforward, but the analysis of the choked version is much more complicated. As a consequence most small feeds are designed empirically with measured data.

4.4.4.3 *Multi-mode conical horns*

These improve the performance of conical horns by generating a second mode inside the horn in such a manner that the aperture fields are linearised. The second, TM_{11} mode, is generated by a step change in the conical horn diameter and the length of the horn is determined by the need to have the modes in the correct phase relationship at the aperture. The dual mode horn gives low cross-polarisation over a narrow band of frequencies. Although narrow band it is simple to make and of low weight.

The concept of adding higher order modes in a horn can be extended for other purposes. In tracking feeds a higher order mode is used to provide tracking information. The inherent cross-polarisation which occurs in offset reflectors can be cancelled by the appropriate addition of higher order modes (Love, 1986). Finally the main beam can be shaped to provide higher efficiency in prime focus reflectors although only over a narrow frequency band.

4.4.4.4 *Conical corrugated horns*

These are the leading choice for a feed for dual reflector and medium size offset reflectors, Figure 4.11(b). They have excellent radiation pattern symmetry and radiate very low cross-polarisation over a broad range of frequencies.

A corrugated horn propagates a mixture of TE_{11} and TM_{11} modes called a hybrid HE_{11} mode. The corrugations are approximately quarter of a wavelength deep so that the electric short circuit at the base of the slot is transformed to a magnetic short circuit at the top of the slot. The result is that the azimuthal magnetic field is forced to zero at the corrugations and the azimuthal electric field is zero due to the ridges. Consequently the boundary conditions of the TE and TM modes are identical and the mutual propagating modes are linear combinations of the two parts. The design procedure for corrugated horns is well understood (Clarricoats, 1984a) and it is possible to accurately predict the radiation characteristics.

4.4.4.5 *Array feeds*

They are used to form multiple beam and shaped beam reflector antennas used on satellites, Figure 4.11(e). The individual elements of the array can be any type of horn, although for compactness small diameter open ended waveguides are preferred. The radiation patterns of the array are mainly determined by the element spacing and the amplitudes and phases of the signals sent to the individual elements. In addition to being able to form a wide range of multiple or shaped beams, the array has the advantage that the cross-polarisation of the total array is lower than that of an individual element. However the closeness of the array elements gives rise to mutual coupling between the aperture fields which can distort the radiation patterns (Clarricoats, 1984b; Clarricoats, 1984c). A significant disadvantage of an array is that a beam forming network of waveguide components must be used behind the array elements to produce the correct amplitudes and phases to the array. For large arrays this can be heavy, expensive and a significant part of the design of the complete antenna system.

4.5 Antennas used in communication systems

4.5.1 Microwave line of sight radio

A typical microwave radio relay system consists of two axi-symmetric parabolic reflector antennas on towers, Figure 4.12, with a spacing of the order of 50km apart in a line of sight path. The relationship between the transmitted and received powers and the antenna and path parameters is given by the Friis transmission formulae, Equation

Figure 4.12 Microwave line of sight reflector antenna

4.2. The typical antenna gain is about 43dBi which means a diameter of about 3 metres at 6GHz.

In addition to the pattern envelope specifications (which must be low because two or more antennas are normally mounted next to each other on a tower), there are a number of other important criteria for microwave radio antennas.

The front-to-back ratio must be high, and the cross-polar discrimination needs to be high for dual polarisation operation, typically better than −25dB within the main beam region over a bandwidth of up to 500MHz. The VSWR needs to be low (typically 1.06 maximum) in a microwave radio relay system in order to reduce the magnitude of the round trip echo.

The supporting structure must be stable to ensure that the reflector does not move significantly in high winds. The reflectors must operate under all weather conditions, which means that a radome is often required. This poses extra design problems because inevitably it degrades the electrical performance. A long waveguide or coaxial cable feeder must be provided from the transmitter to the antennas. Not only must this be low loss but it must also be well made so that there is no possibility of loose joins introducing non-linear effects. Finally the cost must be relatively low because a large number of reflectors are required in a microwave communication system.

The majority of antennas in use are prime focus symmetric reflectors often with shields and radomes. The design of the prime focus reflectors follows the procedure discussed in earlier sections. The need to have a high front-to-back ratio means that either a low edge illumination must be used or baffles and shields must be used. The latter methods are preferable, but increase the weight and cost. The most common feed is a modified TE_{11} circular waveguide, which is designed to have a low VSWR and good pattern symmetry. Sometimes operation in two frequency bands is needed in which case the feed must combine two waveguides and operate at the two frequencies.

The VSWR can be reduced by replacing the centre portion of the paraboloid with a flat plate, called a vertex plate. This minimises the VSWR contribution from the dish although it also degrades the near-in sidelobes and reduces the gain.

4.5.2 Earth station antennas

Earth station antennas are at the earth end of satellite links. High gain is needed to receive the weak signals from the satellite, or to transmit strong signals to the satellite. The antennas can be divided into three types:

1. Large antennas required for transmit and receive on the IN-TELSAT type global networks with gains of 60 to 65dBi (15 to 30 metres diameter).
2. Medium sized antennas for cable head (TVRO) or data receive only terminals (3-7 metres diameter)
3. Small antennas for direct broadcast reception (0.5-2 metres diameter).

Types 1 and 2 have to satisfy stringent specifications imposed by regulatory bodies. When the recommended spacing of satellites in the geostationary arc was 3 degrees, the pattern envelope was specified by $32 - 35 \log\theta$. This could be met with a symmetric reflector antenna. With the new spacing of 2 degrees, the pattern spec has been improved to $29 - 25 \log \theta$, Figure 4.13. This can best be met with low sidelobe, offset reflector designs.

The minimum receivable signal level is set by inherent noise in the system. Earth stations are required to detect small signals so the control of the noise parameters is important. The noise appearing at the output terminals of an earth station used as a receiver has three components; the noise received by the main beam of the reflector; the spillover noise due to the spillover from the feed; the receiver noise. The first component may be due to natural sources or to man made interference. The natural noise emitters are the earth and sea absorption, galactic noise, isotropic background radiation, quantum noise and absorption due to the oxygen and water vapour in the earth's atmosphere. A minimum isotropic background radiation of about 3K is always seen by any antenna. The value of the other factors depends on frequency. The spillover noise is the only component under the control of the antenna designer. Its value can be reduced by designing an antenna with very low sidelobes. The receiver noise is normally

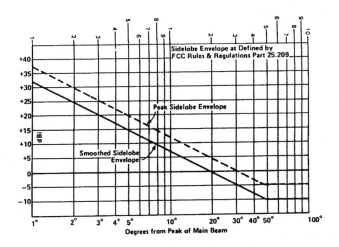

Figure 4.13   FCC earth station antenna pattern specification

the dominant noise factor. It depends on the method of amplification and detection.

Early earth stations all used cooled receivers which have low noise temperatures. Modern earth stations use uncooled receivers which are dependent on the noise performance of the front-end transistor.

This was improved dramatically in recent years, especially for small DBS terminals where the economies of scale have supported considerable research to reduce the noise temperature.

The ratio of the gain to noise temperature, the G/T ratio, is a useful measure of the influence of the noise components. Typical values are $40.7 dBK^{-1}$ for an INTELSAT A, 30 metre diameter antenna operating at 4/6GHz (Pratt, 1986)

Large earth station antennas are expensive to construct and to maintain so that there is a premium in obtaining the maximum efficiency from the system.

The axi-symmetric Cassegrain antenna (see Section 4.4.2) is the favourite choice for a number of reasons:

1.   The gain can be increased over the standard parabola hyper-
        bola combination by shaping the reflectors. Up to an extra 1dB
        is possible.
2.   Low antenna noise temperatures can be achieved by control-
        ling spillover using a high performance corrugated horn and
        by using a beam waveguide feed system.
3.   Beam waveguide feed systems place the low noise receivers
        and high power transmitters in a convenient, stationary, loca-
        tion on the ground.

The beam waveguide feed system (Rudge, 1986), Figure 4.14, consists of at least four reflectors, whose shape and orientation is chosen so that the transmitter and receiver can be stationary whilst the antenna is free to move in two planes. The free-space beam suffers very little loss. The dual polarised transmit and receive signals need to be separated by a beam-forming network placed behind the main feed horn. For 4/6GHz operation, this will also incorporate circular polarisers.

The narrow beam from the large antenna necessitates the incorpor-ation of some form of tracking into the antenna because even a geostationary satellite drifts periodically. There are a number of schemes available, including monopulse, conical scan and hill climb-ing. The favourite is a monopulse scheme using additional modes in the feed horn to electromagnetically abstract the tracking data.

The first generation of medium earth station antennas were axi-symmetric Cassegrain reflector antennas, sometimes shaped. How-ever the advent of tighter pattern specifications has led to the wide-spread use of single or dual offset reflector antennas (see Section 4.4.3). These can meet the low sidelobe specifications by removing blockage effects from the aperture. Very high efficiency designs have been produced by shaping the reflectors to optimise the use of the aperture (Bergman, 1988; Cha, 1983; Bjontagaard, 1983). For these high efficiency designs, the r.m.s. surface error on the main reflector needs to be less than 0.5mm for operation in the 11GHz to 14GHz band. The feed is a high performance corrugated horn. The offset reflector configuration lends itself to deployment and portable de-

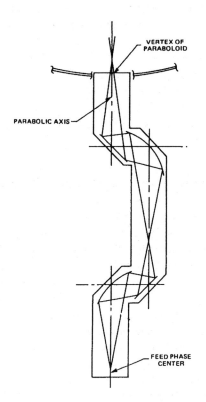

Figure 4.14   Beam waveguide feed system for large earth stations

signs have been produced where the offset reflector folds for trans-
portation.

Cost is the main driver for small earth station antennas for mass
market applications. Receive only terminals in the 4/6GHz band for
data or TV reception are usually symmetric prime focus paraboloid
which are made by spinning an aluminium sheet. In the 11GHz
communication band or the 12GHz DBS band, prime focus offset

reflectors made from fibreglass moulds are popular. A simple open-ended waveguide type feed is incorporated on a sturdy feed support with the first stage low noise converter incorporated directly into the feed. There is considerable interest in making flat-plate array antennas which can be mounted flush against buildings and incorporate electronic scanning to look onto the satellite signals. The technology for electronic scanning is available from military radars but not so far at a price which is acceptable to the domestic market.

4.5.3 Satellite antennas

4.5.3.1 *Telemetry, tracking and command (TT&C)*

The ideal TT&C antenna would give omnidirectional coverage so that the orientation of the satellite would be irrelevant. Wire antennas are used for VHF and UHF coverage but the space craft is a few wavelengths across at these frequencies and therefore considerable interaction between the antenna and the satellite distorts the radiation pattern. An alternative approach is to use a low gain horn antenna to provide full earth coverage. This is particularly useful for spin stabilised spacecraft. The earth subtends 17 degrees from a geostationary satellite which can be met with a small conical horn.

4.5.3.2 *Spot beams*

Spot beam antennas are required to produce a beam covering a small region of the earth's surface. The angular width of the beam is inversely proportional to the diameter of the antenna. Size considerations virtually dictate that some form of deployable mechanism is needed on the satellite and this leads to the use of offset reflectors with a dual-mode or corrugated feed horn. The constraints of the launcher mean that the maximum size for a solid reflector is about 3.5 metres. Larger reflectors can only be launched by using some form of unfurlable mesh or panel reflector. The trend towards smaller footprints on the earth can be met either by using a larger reflector or by using a higher frequency, both of which involve higher costs. To date most spot beam communication satellites have used two prime focus offset reflectors, one for transmit and one for receive, producing footprints on the earth's surface which are elliptical because of the curvature of the earth.

4.5.3.3 *Multiple beams*

It was early recognised that by using a single reflector and an array of feeds it was possible to produce multiple beams on the earth, Figure 4.15. This has the advantage that most of the antenna sub-system is re-used with the penalty of having to design and make the array of feed horns and the beam forming network behind the array. The array feed elements must be compact so that they occupy the minimum space in the focal plane of the offset reflector. At the same time the cross-polarisation must be low. This tends to mean that corrugated horns cannot be used and small diameter dual-mode rectangular or circular horns are preferred. The maximum number of beams depends on the tolerable aberrations since array elements which are off-axis will have degraded performance.

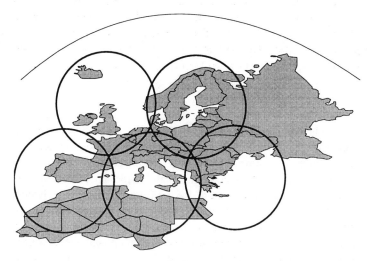

Figure 4.15 Multiple spot beams generated by an array feed on a space craft

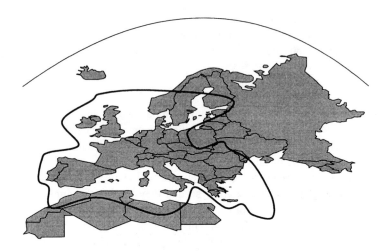

Figure 4.16   European contoured beam

4.5.3.4 *Shaped beams*

It is desirable to optimise the shape of the satellite beam on the earth's surface so as to conserve power and not waste energy by illuminating portions of the oceans. An example is shown in Figure 4.16. Shaped beams can be produced in two ways. Multiple, overlapping beams produced by an offset parabolic reflector and an array of feeds can be used. This approach is an extension of the multiple beams and has the advantage that it is possible to design for reconfiguration by incorporating switching systems into the beam forming network. The alternative approach is to use a single, high performance feed and to physically shape the surface of the reflector so that power is distributed uniformly over a shaped beam region. Both approaches have received considerable attention in recent years.

The multiple beam approach is well illustrated by the INTELSAT VI communication satellite, Figure 4.17, which produces multiple shaped beams to cover the main population regions of the earth. (Figure 4.18.) In order to be able to use the same satellites over the Atlantic, Indian or Pacific Oceans, the array feed consists of 146

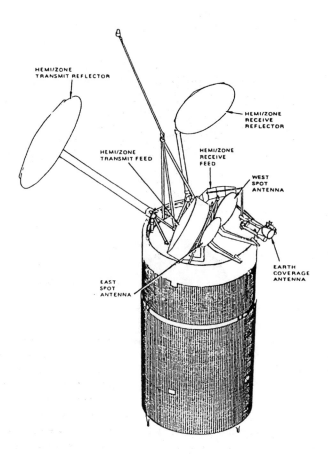

Figure 4.17 INTELSAT IV antenna system

elements which can be switched to produce the appropriate shaped beams (Bennett, 1984).

The shaped reflector approach has the advantage of mechanical simplicity and lower weight at the penalty of fixed beams. The

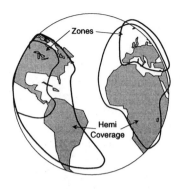

ATLANTIC/PACIFIC OCEAN

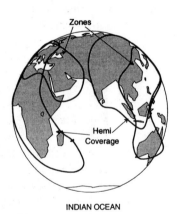

INDIAN OCEAN

Figure 4.18 Shaped beam generated by INTELSAT VI antenna

theoretical design process is quite extensive and involves a synthesis process with the input of the required beam shape and the output of the contours of the reflector surface.

A single offset reflector constrains the possible shapes because it is not possible to arbitrarily specify the amplitude and the phase of the synthesised pattern. This constraint is removed with a dual reflector design.

4.5.4   VHF and UHF communications

Antennas for VHF and UHF communication systems take on a wide variety of specific forms, but the vast majority are derivatives of the generic dipole type antenna. The physical, mechanical and environmental aspects are generally more significant than for microwave antennas because the smaller size of the antenna means that the radiation and impedance characteristics are partly determined by these aspects.

A comprehensive survey of VHF and UHF antennas can be found in (Rudge, 1986; Johnson, 1984). Antennas that give near uniform coverage in one plane can be obtained from half wave dipoles or monopoles. Complementary antennas such as loops and slots will work equally well and the actual shape will be determined more by the application than by the basic electromagnetic performance. The bandwidth of these simple elements is limited by the impedance characteristics, although most communication applications only require relatively narrow bandwidths. With small elements, some form of impedance matching network is required. One problem with balanced dipole type antennas is that they are required to be fed by an unbalanced coaxial cable. A balun is needed to match the balanced to unbalanced system and this is inevitably frequency sensitive.

Antennas for point-to-point links need to be directional and have as high a gain as possible. This is achieved with Yagi-Uda array, Figure 4.19, which consists of one driven element, one reflector element and a number of director elements. Only the driven element is connected to the feed line; the other elements are passive and currents are induced in them by mutual coupling, the spacing ensuring that this is in the correct amplitude and phase to give a directional radiation pattern. Gains of up to about 17 dBi are possible from one Yagi-Uda array. Higher gains can be obtained by multiple arrays. The Yagi-Uda array is inherently linearly polarised. Circular polarised arrays can be made either from crossed dipoles or from helixes.

Antennas for mobile communications can be divided into those for base stations and those for the mobiles. Base station antennas are mounted on towers and usually require to have nearly uniform patterns in the horizontal plane with shaping in the vertical plane to

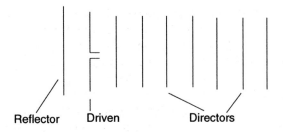

Figure 4.19   A Yagi-Uda array

conserve power. This can be achieved with a vertical array of vertical dipoles or other panelled dipoles. The influence of the tower on the antenna must be taken into account in the design.

Mobile antennas on vehicles, ships, aircraft or near humans present challenging problems to the antenna designer. In most cases the physical, mechanical and environmental aspects take precedence over the electromagnetic design. In consequence the ingenuity of the antenna designer is required to produce an antenna which works well in adverse conditions. For instance antennas on aircraft must not disturb the aerodynamic profile so cannot protrude from the body of the aircraft. The effects of corrosion, temperature, pressure, vibration and weather are other factors to be taken into account. Antennas for personal radios are constrained by the role of the operator and by the need for very compact designs commensurate with satisfying radiation safety levels. The human body acts partly as a director and partly as a reflector depending on the frequency of use and the relative position of the antenna to the body. The portable radio equipment has to be considered a part of the antenna system including the radio circuits, batteries and case. In general, improved performance will result when the antenna is held as far from the body as possible and as high as possible.

4.5.5 HF communications

HF antennas are used in the range of frequencies from 2 to 30MHz for mobile communications and some fixed communications. Space precludes more than a brief mention of the types of HF antennas. Surveys can be found in Rudge (1986) and Johnson (1984). HF antenna design is constrained by the ionospheric propagation characteristics which change both daily, seasonally and with the sun-spot cycle. Antennas can receive either the sky wave reflected from the ionosphere or the ground wave if transmitter and receiver are close together. The wavelength in the HF band is such that antennas are usually only a fraction of a wavelength in size and this in turn means that the local environment around the antenna will have a major impact upon the performance. This is particularly true for antennas mounted on vehicles, ships and aircraft. The analysis of the antenna must take account of the environment by techniques such as wire grid modelling (Mittra, 1975). This is computer intensive and inevitably approximate which means that much HF antenna design is empirical.

Most HF antennas are based on dipoles, monopoles or wire antennas. Complementary elements such as loops or slot antennas are also used. Directionality or gain is achieved by arrays of elements. A prime requirement of most HF antennas is that they are broadband in order that the optimum propagation frequency can be used. The radiation patterns of the basic elements are wide band but the input impedance or VSWR is narrow band. To overcome this limitation a tuning unit has to be incorporated into the system. The wide band operation is achieved with automatic tuning units.

4.6 References

Balanis, C.A. (1982) *Antenna Theory: Analysis and Design* Harper & Row, New York.

Barrett, M. and Arnott, R. (1994) Adaptive antennas for mobile communications, *Electronics & Communication Engineering Journal*, August.

Bennett, S.B. and Braverman, D.J. (1984) INTELSAT VI - A Continuing Evolution *Proc. IEEE* , **72**, 1457.

Bergman, J., Brown, R.C., Clarricoats, P.J.B. and Zhou, H. (1988) Sythesis of Shaped Beam Reflector Antenna Patterns, *Proc. IEE*, **135 (H)**, (1) 48.

Bjontagaard, G. and Pettersen, T. (1983) An Offset Dual Reflector Antenna Shaped from Near-Field Measurements of the Feed Horn; Theoretical Calculations and Measurements, *IEEE Trans.*, **AP- 31**, 973.

Browne, J. (1995) Advances in antennas drive wireless systems, *Microwaves & RF*, May.

Cha, A.G. (1983) An Offset Dual Shaped Reflector with 84.5 Percent Efficiency, *IEEE Trans.*, **AP-31**, 896.

Clarricoats, P.J.B. and Olver, A.D. (1984a) *Corrugated Horns for Microwave Antennas*, Peter Peregrinus (IEE), London.

Clarricoats, P.J.B., Tun, S.M. and Parini, C.G. (1984b) Effects of Mutual Coupling in Conical Horn Arrays, *Proc. IEE*, **131 (H)**, (1), 165.

Clarricoats, P.J.B., Tun, S.M. and Brown, R.C. (1984c) Performance of Offset Reflector Antennas with Array Feeds, *Proc IEE*, **131 (H)**, (1) 172.

Collin, R.E. and Zucher, F.J. (1969) *Antenna Theory (Pts 1 & 11)*, McGraw Hill, New York.

Duret, G. (1994) Antennas for communication satellites, *Electrical Communication*, 4th Quarter.

Elliot, R.S. (1981) *Antenna Theory and Design*, Prentice Hall, Englewood Cliffs, N.J.

Foster, P.R. (1994) Antenna installed performance using diffraction theory, *Electronics & Communication Engineering Journal*, October.

Galindo, V. (1964) Design of Dual Reflector Antennas with Arbitrary Phase and Amplitude Distribution, *IEEE Trans.*, **AP- 12**, 403.

Hubermark, S. (1993) Antenna design parameters in mobile comms, *Mobile Europe*, December.

James, G.L. (1986) *Geometrical Theory of Diffraction for Electromagnetic Waves*, Peter Peregrinus (IEE), London.

Johnson, R.C. and Jasik, H. (Eds) (1984) *Antenna Engineering Handbook*, McGraw-Hill, New York.

King, R.W.P. (1956) *The Theory of Linear Antennas* , Havard Univ. Press. Cambridge, Mass.

King, R.W.P., Mack, R.B. and Sandler, S.S. (1968) *Arrays of Cylindrical Dipoles*, Cambridge Univ. Press, New York.

Lamb, J.W. and Olver, A.D. (1986) Blockage Due to Subreflector Supports in Large Radiotelescope Antennas, *Proc. IEE*, **133(H)** 43 (1).

Lo, Y.T. and Lee, S.W. (1988) *Antenna Handbook*, Van Nostrand, New York.

Love, A.W. (1976) *Electromagnetic Horn Antennas*, Selected Reprint Series, IEEE Press, New York.

Love, A.W., Rudge, A.W. and Olver, A.D. (1986) Primary Feed Antennas. *In The Handbook of Antenna Design, (eds. A.W. Rudge et al)*, Peter Peregrinus (IEE), London.

Ludwig, A.G. (1973) The Definition of Cross- Polarisation, *IEEE Trans.*, **AP-211** (1), 116.

Milligan, T. (1985) *Modern Antenna Design*, McGraw-Hill, New York.

Mittra, R. (1975) *Numerical and Asymptotic Methods in Electromagnetics*, Springer Verlag, Berlin.

Nowicki, D. and Roumellotis, J. (1995) Smart antenna strategies, *Mobile Communications International*, April.

Olver, A.D. (1986) Basic Properties of Antennas. *In The Handbook of Antenna Design (eds. A.W. Rudge, et al)* Peter Peregrinus (IEE), Chapter 1.

Pratt, T. and Bostian, C.W. (1986) *Satellite Communications*, Wiley, New York.

Rahmat-Samii, Y. (1986) Reflector Antennas. *In Antenna Handbook (eds. Y.T. Lo and S.W. Lee)* Van Nostrand, New York. Chapter 15.

Rudge, A.W., Milne, K. Olver, A.D. and Knight, P. (Eds.) (1986) *The Handbook of Antenna Design*, Peter Peregrinus (IEE), London.

Rusch, W.V.T. and Potter, P.D. (1970) *Analysis of Reflector Antennas*, Academic Press, New York.

Rusch, W.V.T., Ludwig, A.C. and Wong, W.G. (1986) Analytical Techniques for Quasi-Optical Antennas. *In The Handbook of Antenna Design (eds. A.W. Rudge, et al)* Peter Peregrinus (IEE). Chapter 2.

Schneiderman, R. (1995) Antenna makers set 'smart' goals, *Micorwave & RF*, May.

Shelton, S. (1994) An update on glass mount GSM antennas, *Global Communications*, May/June.

Silver, S. (Ed.) (1984) *Microwave Antenna Theory and Design*, Peter Peregrinus (IEE), London.

Stutzman, W.L. and Thiele, G.A. (1981) *Antenna Theory and Design*, Wiley, New York.

Warwick, M. (1993) Antennas: a basic guide to requirements, *Communications International*, January.

Williams, W.F. (1965) High Efficiency Antenna Reflector, *Microwave J.*, **8**. 79.

5. Line of sight radio systems

5.1 Microwave system path design considerations

A thorough understanding of the principles of microwave path design is essential in the planning and engineering activities for a microwave system. The operator of a microwave system should also understand these principles if he is to evaluate the work of an engineering organisation designing his system.

5.1.1 Path clearance

Microwave energy tends to travel in a straight line between microwave stations — hopefully, most of the time. Unfortunately, the system designer is faced with the unavoidable fact that we live on a round earth. Thus, the curvature of the earth's surface between adjacent microwave stations must be taken into account when determining the antenna heights. The formula for calculating earth curvature is as in Equation 5.1 where h is the earth's curvature in feet, d_1 and d_2 in miles, and K is a constant.

$$h = \frac{d_1 \, d_2}{1.5 \, K} \tag{5.1}$$

In this equation d_1 and d_2 are the distances from the path location in question to either end of the microwave hop.

As may be seen from Figure 5.1, earth curvature has its greatest effect upon path clearance at the centre of the path. However, when calculating the total path clearance required using additional formulas to be described later, the earth curvature must always be included regardless of the location along the path.

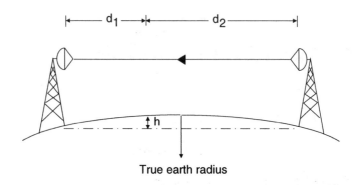

Figure 5.1   Effect of earth curvature on path clearance

The K factor in Equation 5.1 is a constant whose value depends upon the actual propagation of the microwave energy along the path. Various values of K are used to describe propagation that differs from a straight line, and an understanding of these 'K factors', as they are called, is important.

Microwave energy, due to its very high frequency (and corresponding short wavelength) tends to behave much like visible light. Microwave energy can be focused, reflected, and refracted (or bent) by the atmosphere. To describe an atmospheric effect that bends the microwave energy away from a straight line, the K factor that appears in the earth curvature formula is changed. The K factor used to describe straight line microwave propagation is 1.0.

Because the atmospheric density decreases as the height above the earth increases, the lower part of the wavefront of the microwave signal tends to travel slightly slower than the upper part of the wavefront. This results in a slight downward curvature of the signal (Figure 5.2). For this case a K factor of 4/3 is used, meaning that if the microwave path is drawn as a straight line, the earth will appear flatter than normal, or will have a radius greater than normal. A path design using 4/3 as the K factor would result in less lower height on each end of the path. However, a conservative design approach will use a normal K factor of 1.0.

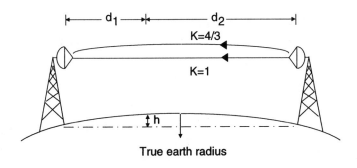

Figure 5.2   'Normal' 4/3 propagation

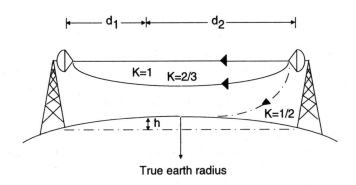

Figure 5.3   Earth bulge effect

The possibility of abnormal propagation must also be considered. An effect known as earth bulge occurs more often than a system designer would like, especially in humid coastal areas. An atmospheric condition such as an inversion layer, with atmospheric density increasing with elevation will cause a bending of the microwave signal opposite to the curvature of the earth and, if the inverse bending is severe, the microwave signal will be diverted into the earth (Figure 5.3). The term 'earth bulge' arises from the apparent bulging of the earth sufficiently to block the microwave signal. The K factor

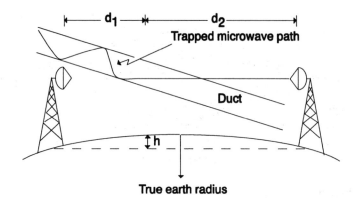

Figure 5.4   Atmospheric duct effect

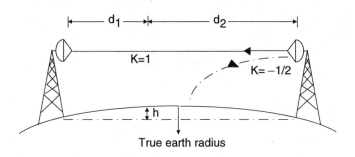

Figure 5.5   Super refractive effect

most often used to protect against outage due to earth bulge is 2/3, although very conservative designs may use factors as low as 1/2.

Other propagation anomalies may occur from time to time. These include ducting i.e. steering of the microwave signal away from the receiving antenna by atmospheric ducts, Figure 5.4; super refractive effects which direct the signal into the earth, Figure 5.5; decoupling effects which cause the signal at the receiving antenna to arrive

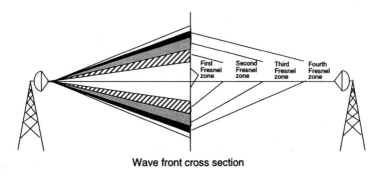

Wave front cross section

Figure 5.6   Fresnel zone concept

outside the main lobe of the antenna pattern and reflective effects from atmospheric 'sheets' causing cancellation of the main microwave signal. It is beyond the scope of this discussion to elaborate on these effects which, although relatively rare, should interest the serious microwave designer.

At this point the reader might be tempted to conclude that as long as earth curvature is taken into account, or as long as line of sight exists between the two ends of a microwave path, that the path clearance is satisfactory. This is not the case.

As microwave energy propagates from the transmitting antenna, an expanding wavefront is created. Looking at the wavefront on a cross-section basis, it may be seen to consist of a series of concentric areas around the path line as shown in Figure 5.6. These areas are numbered beginning with the area immediately around the path line. The signal at the receiving antenna will be affected if a portion of the wavefront is blocked, or if a portion of the wavefront is reflected from an obstruction along the path.

The first Fresnel zone consists of the first area around the path line. A path designed so that only this area is not blocked by any obstructions below the path would be described as having a first Fresnel zone clearance above the earth.

Another important consideration is the nature of the earth along the path. A reflective surface such as water may actually result in partial

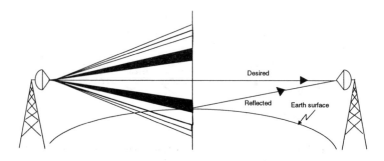

Figure 5.7   Effect of second Fresnel zone clearance

or complete cancellation of the desired signal. For example, if the earth is perfectly reflective (R = –1.0), the desired signal will be cancelled with a path clearance equal to 2F as in Figure 5.7. Even with a less reflective earth (and a typical reflection coefficient is – 0.3) a partial cancellation may occur if the clearance is equal to any even Fresnel zone number and the obstruction is reflective. 'More is better' is not necessarily the case with path clearance.

The formula for calculating Fresnel zone clearance is as Equation 5.2 where F_1 is the distance in feet from the path line to the edge of the first Fresnel zone; d_1 and d_2 are the distances in miles from the point in question to each end of the path; D is the total path length in miles; and f is the frequency in GHz.

$$F_1 = 72.2 \left(\frac{d_1 \, d_2}{D f} \right)^{\frac{1}{2}} \tag{5.2}$$

Fresnel zone clearances greater than F_1 may be calculated by Equation 5.3 where N is the Fresnel zone number.

$$F_N = F_1 \sqrt{N} \tag{5.3}$$

Propagation anomalies that alter the path line will obviously change the clearance above obstructions. It is possible that received signal variations may occur due to the clearance equalling an even numbered Fresnel zone clearance. This, in combination with a highly reflective surface, could cause a deep fade in the received signal.

It is worth noting that if earth bulge using a $K = \frac{2}{3}$ is considered, many path designers allow a zero Fresnel zone clearance (also called grazing).

5.1.2 Path profiles

The first step in planning a microwave path is to examine topographic maps along the path line for potential obstruction points. Obstructions to the sides of the path line such as cliffs or buildings should also be noted, especially if they can be considered reflective. A path profile may then be plotted showing the elevations of the path ends as well as elevations of points along the path.

Several methods of plotting this information have been used. Figure 5.8 illustrates these methods. Use of a curved base line to represent each curvature or use of a curved template to draw the path line over a flat earth base line are two methods. With the advent of the modern scientific calculator, the easiest method is to use a straight earth line and a straight path line with the necessary clearance over each obstruction calculated and shown by a symbol. Both the earth curvature and the desired Fresnel clearance must be calculated and added in order to plot the clearance target above each obstruction.

5.1.3 Field path survey

In most cases, the path design derived from the map study must be confirmed by a field survey. Investigation of the potential microwave sites is usually necessary, and information on man-made obstructions as well as confirmation of natural obstructions should be obtained. Potential reflection points may also be determined.

A microwave path designed and installed without the benefit of a field survey is a high risk except for the obvious types of paths where the path characteristics are obvious and the path is short. A typical microwave survey report generated by the survey contractor will contain most of the site and path data needed for FCC licence applications and FAA approval in the case of towers near an airport.

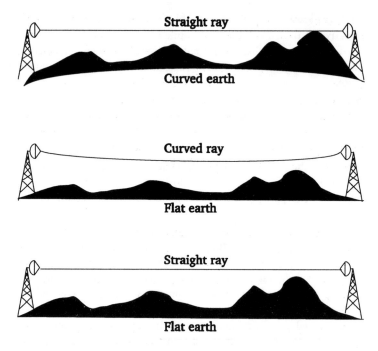

Figure 5.8 Three methods of drawing path profiles

In the case of potential reflection points observed along the path, it is possible to design a path so as to minimise the possibility of signal cancellation by a reflected signal. Calculation of the potential reflection areas that would cause cancellation may be made in advance of the actual field work, minimising the need for extensive field investigation along the path.

5.1.4 Propagation

Once the physical characteristics of a microwave path have been determined we can turn our attention to the determination of antenna systems and propagation availability (reliability) that might be expected.

Calculation of the path loss, or attenuation of the microwave signal between the ends of the path is the first step. In order to make the formula as generally applicable as possible, the path loss calculation assumes the use of a theoretical point-source or isotropic antenna at each end of the path. With this concept the microwave energy is radiated uniformly in all directions at the transmitting end and the receiving antenna provides no directivity. The loss formula is given by Equation 5.4 where loss is expressed in dB; f is the frequency in GHz; D is the path length in miles. For example the path loss at 6.7GHz for a 25 mile path would be 141dB.

$$Loss = 96.6 + 20 \log_{10} f + 20 \log_{10} D \qquad (5.4)$$

Fortunately, microwave energy can be focused and reflected using a parabolic antenna at each end of the path. The gain or focusing effect of a typical antenna is 25dB to 45dB over an isotropic antenna, depending on the size of the antenna and the operating frequency. Thus with reasonable transmitter power and receiver sensitivity, satisfactory transmission may be achieved.

We may now determine the signal at the microwave receiver input for a typical path. Figure 5.9 is a graphical representation of the factors to be considered in this calculation. At the left side of the figure, the transmitter output power is shown. The loss of the transmission line is shown as a decreasing signal between the transmitter output and the input to the antenna. The antenna provides a gain, shown as a signal increase. The path loss, representing the largest single contributor, is shown as a decreasing signal as the path is traversed.

Similarly, the antenna gain and transmission line loss are shown at the receiver site. Notice that the signal at the receiver input is significantly higher than the receiver sensitivity, or minimum acceptable receiver input signal. This is intentional and the difference may be as large as 40dB or 50dB. The difference is known as fade margin.

Microwave signals, even under normal atmospheric conditions where none of the previously mentioned phenomena such as earth bulge, ducting, reflections, etc., occur, may still be subject to variation due to an effect called multipath fading. Multipath fading occurs

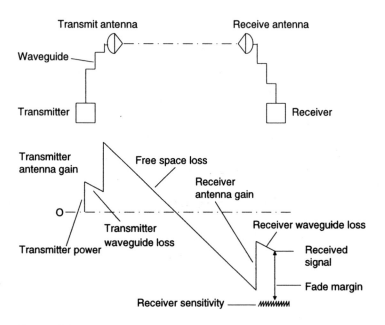

Figure 5.9   Path calculations

when the microwave energy is diffracted or scattered along the path so that part of the signal arrives at the receiving antenna out of phase with the desired signal. Figure 5.10 shows this effect with the phases of the main and scattered signals compared.

Mathematically, multipath fading is described by the Rayleigh distribution function. Figure 5.11 shows this relationship. Notice that as the permissible fade depth is increased, the probability of cancellation decreases.

Now the reason for fade margin should be apparent. With a path designed for a 40dB fade margin, the probability of a fade deep enough to reach the receiver sensitivity point and cause the receiver to squelch is 0.01%. This corresponds to a path availability (or reliability) of 99.99%, or an outage time of 53 minutes per year. For a system consisting of a single microwave hop, this may seem acceptable. Consider, however, a system of 20 tandem hops, each

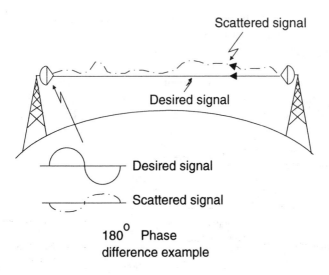

Figure 5.10   Multipath effect

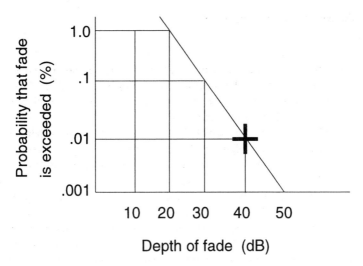

Figure 5.11   Raleigh fading

with a 53 minute outage per year. Murphy's law says that no two paths will fade simultaneously so that a propagation outage time of 20 times 53 minutes per year, or 17.67 hours per year, might be expected. This would be unacceptable to most system users.

About 20 years ago W.T. Barnett and A. Vigants of Bell Telephone Laboratories developed new methods of calculating propagation unavailability based upon extensive experimental and theoretical work. These methods take into account the frequency and path length as well as climate and topographic factors. The formula for propagation unavailability (outage) is given as in Equation 5.5.

$$U = (a)(b)(2.5 \times 10^{-6})(f)(D^3)(10^{-F/10}) \tag{5.5}$$

U is the unavailability in decimal form. a is a topographic factor and equals: 4 for very smooth terrain including water; 1 for average terrain with some roughness; 0.25 for mountains, rough or very dry terrain. b is a climate factor and equals: 0.5 for Gulf coast or other hot humid locations; 0.25 for temperate or northern areas; 0.125 for mountainous or very dry areas. f is the frequency in GHz. D is the path length in miles. F is the fade margin in dB.

The formula assumes, of course, that the path has been designed for adequate clearance and does not take into account effects such as earth bulge, ducting, reflections, etc. The decimal form of U may be converted into time by multiplying by the number of minutes (or seconds or hours) per year. This formula and the unavailability formulas following apply to one-way propagation outages. The unavailability time should be doubled for a complete two-way path, since a conservative design would assume that outages in both directions do not occur simultaneously.

Figure 5.12 illustrates the propagation unavailability for several typical length 6GHz paths as compared to the Rayleigh distribution function.

At this point it may seem that the system designer, if faced with an unacceptable outage time for the path, must consider increasing fade margin, changing to a lower frequency microwave band or shortening the path in order to decrease the propagation outage time. Another solution is available.

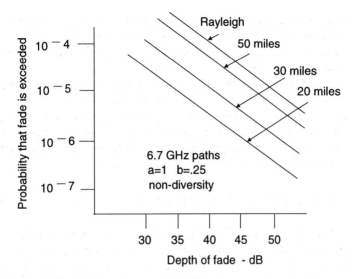

Figure 5.12 Propagation unavailability

Consider the situation shown in Figure 5.13. A single transmitting antenna is used, and the two receiving antennas are usually spaced vertically 30 to 60 feet. Each receiving antenna is connected to a separate receiver and the receiver baseband outputs are then connected to provide a combined signal (Figure 5.14). The desired microwave signal is received by both antennas, but should a cancellation at one antenna occur due to multipath or a reflected signal, a simultaneous cancellation on the other antenna is not likely, due to the different path length of the reflected signal. The formula for calculating the improvement to be expected by using space diversity is as in Equation 5.6, where I_{SD} is in decimal form; S is the vertical antenna spacing in feet; F is the lower fade margin associated with the receiving antenna with the longer transmission line.

$$I_{SD} = \frac{(7 \times 10^{-5})(f)(S^2)(10^{F/10})}{D} \qquad (5.6)$$

Typical values of I_{SD} are in the range of 100 to 500.

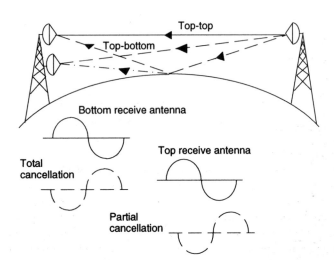

Figure 5.13   Space diversity

The formulas for U and I_{SD} may be combined to provide an unavailability formula for the space diversity case as in Equation 5.7 where U_{SD} is the unavailability in decimal form for space diversity. The other factors are as defined previously.

$$U_{SD} = \frac{(a)(b)(3.6 \times 10^{-2})(D^4)(10^{-2F/10})}{S^2} \qquad (5.7)$$

Notice that for U_{SD}, the frequency does not appear in the formula. This indicates that the old 'rule of thumb' requiring greater antenna spacing at lower frequencies is not the case.

Experience indicates that antenna spacing between 30 and 60 feet provides the best compromise between improvement in propagation reliability and increase in cost due to increased tower height. It should be remembered that both the top-to-top antenna path and the top-to-bottom path must meet the necessary path clearance criteria, however.

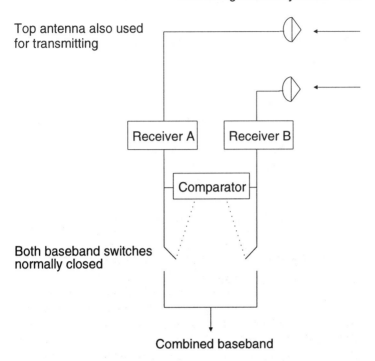

Top antenna also used
for transmitting

Receiver A

Receiver B

Comparator

Both baseband switches
normally closed

Combined baseband

Figure 5.14   Space diversity receivers

While it is theoretically possible to achieve some improvement in propagation availability by mounting the receive antennas 'edge to edge' where space is limited, little experience is available to demonstrate that the formulas are valid for small spacings. The general feeling is that such a configuration would provide an improvement over non diversity operation, however.

As stated, space diversity is an effective method of improving propagation reliability where multipath or reflective type of fading is expected.

Diversity will not provide improvement where earth bulge, ducting, or similar effects that block or steer the signal away from the receiving antenna are present.

5.2 Short haul and millimetre path design

Microwave path design at 18GHz and higher follows a very similar procedure to path design at 2GHz and 6GHz. However, due to the high frequencies involved, some of the parameters evaluated along the way may be quite different than at the lower frequencies. To demonstrate this we have selected a few path design topics and evaluate their significance at 18GHz and 23GHz.

5.2.1 Fresnel zones and obstructions

One of the considerations in microwave path design is the effect of reflections and obstructions of the microwave path.

The effects of reflections and obstructions usually get sufficient attention at the lower frequencies since the paths are much longer and clearances are costly to achieve. It will be noted that the Fresnel zones are larger at the lower frequencies.

At the higher frequencies such as 18GHz and 23GHz the paths are much shorter and it is much simpler to establish line of sight since both ends of the paths usually can be seen.

Figure 5.15 shows a polar plot of an antenna pattern. The main lobe and the side lobes are shown. The beamwidth is shown at the 70 percent voltage point which corresponds to the half power point. The beamwidth is given in degrees. It will be noted that as the wavefront gets farther from the antenna, the dimensions of the wavefront get larger at the half power point. The effects of this will be shown in an example later.

A pictorial of the Fresnel zones is shown in Figure 5.16. Figure 5.16(a) is a cross section of Figure 5.16(b) taken through the beam axis. The rings, both shaded and non-shaded, are known as the Fresnel zones. The distance from T to a point on the circle to R is longer by some multiple of a half wavelength than the main beam. This difference in length is the cause of the Fresnel interference phenomenon.

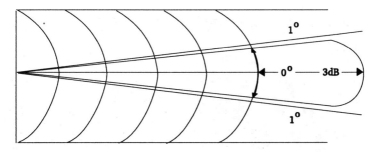

Figure 5.15   Antenna beam width

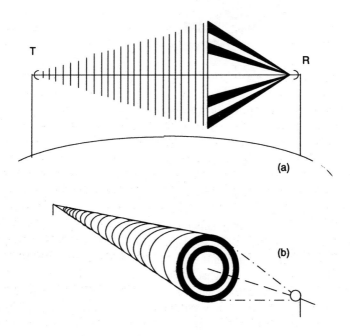

Figure 5.16   Pictorial of Fresnel zones: (a) cross-section of (b) through the beam axis

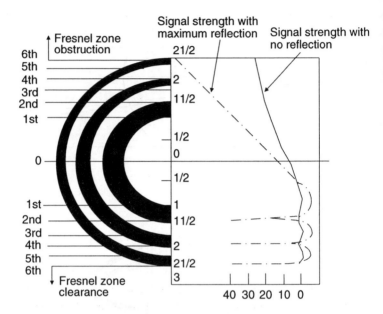

Figure 5.17   DB from free space

The effects at the receiving antenna of having multiple unobstructed Fresnel zones are shown in Figure 5.17. It will be noted that the nulls correspond to the even numbered zones while the peaks correspond to odd numbered zones. It can be seen that with a large number of zones unobstructed and a low reflective surface in the path that the path attenuation is equivalent to free space loss.

As an indication of what happens when various zones are blocked refer to Figures 5.16(a) and 5.17. If the towers of Figure 5.16(a) are lowered, portions of the wavefront will be obstructed by the earth's curvature or an obstacle in the path. As cancelling zones are obstructed the signal will increase and as phase zones are obstructed the signal will decrease. When the obstruction reaches the edge of the first Fresnel zone, the signal will be twice that in free space. Raising or lowering the towers at this point will decrease the signal.

The peaks and valleys will depend on the type of obstruction. A highly reflective obstacle will give deep nulls. If the towers are high enough to clear many zones then an obstacle in the right place can cause deep cancellations.

Some possible solutions to the cancellation problem are:

1.   The Fresnel zones are closely spaced. This indicates that a slight increase or decrease in the antenna heights will remove the cancellation. This can be done if the reflection point is constant and does not change.
2.   If the antenna is increased in size the midpoint illumination is decreased. With a narrower beam antenna the power drops off sharply.
3.   Another method of reducing the amount of reflection is to tilt each antenna up. This can be done if the loss in power can be tolerated.

In conclusion, it can be seen that 23GHz paths may not always be simply 'aim and shoot'. The design of these paths should be done as carefully as the lower frequency paths. Fresnel zones and reflection point should be carefully considered.

5.2.2 Outage considerations

The main reason for being concerned with path design to begin with is to be sure the path availability will meet the requirements placed on the overall system. Assuming the path has been designed for the proper Fresnel clearance etc. the main contributors to outage become multipath fading and rain induced fading. For analogue radios above 10GHz the predominant effect is due to rain induced fading. However, due to the high data rates used, digital radios at these frequencies may be limited by dispersion caused by multipath fades.

5.2.2.1 *Dispersive fading*

One mechanism that causes multipath fading is refraction in the atmosphere. Certain atmospheric conditions cause the received signal

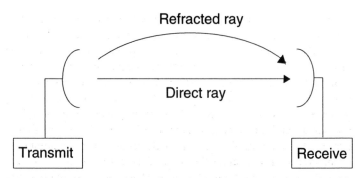

Figure 5.18   Multipath fading

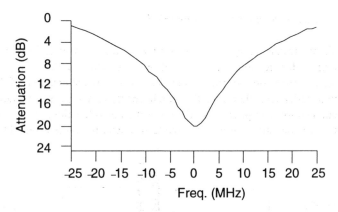

Figure 5.19   Selective fading attenuation characteristic (20dB notch)

to arrive via many separate paths. Figure 5.18 is a simplified representation of multipath fading where only two rays are shown. Since the two rays follow different paths, the distance travelled is different and the two signals may be out of phase. The result is a frequency dependent attenuation characteristic like that in Figure 5.19. For this reason multipath fading is often referred to as frequency selective, or dispersive fading.

The exact shape of the attenuation notch depends on the relative delay and attenuation of the two paths. The difference in level between the two rays determines how deep the notch is while the difference in phase determines the frequency where the notch occurs.

In addition to causing a notch in the attenuation characteristic, multipath causes distortion in the differential delay characteristic. When the delayed ray's amplitude is smaller than the main ray's, a notch in the differential delay occurs. This is referred to as a minimum phase fade. A non-minimum phase fade occurs when the amplitude of the delayed ray is larger than that of the main ray. In this case a 'bump' occurs in the differential delay characteristic while the attenuation characteristic still has a notch in it.

5.2.2.2 *Signature curves and DFM*

To determine the effect of dispersive fading on a digital radio, a multipath notch must be simulated. A test system like that of Figure 5.20 is often used for the simulation. The relative attenuation of the two paths, and therefore the notch depth, is set by the variable attenuators. The phase, and therefore the notch location, is set with the variable phase shifter. The fixed delay determines the width of the

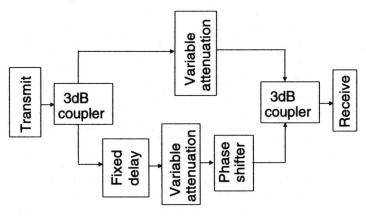

Figure 5.20   Simulation of a multipath notch

notch and is normally set at 6.3ns. The 6.3ns delay is considered statistically representative of most multipath fades.

The measurement is performed by setting the centre frequency of the notch somewhere in the RF channel of the particular radio system under test. The notch depth is adjusted until the Bit Error Ratio (BER) of the radio degrades to 10^{-3}. The phase shifter is then adjusted so that the notch appears at a different frequency and again the notch depth is adjusted for a 10^{-3} BER. This process is repeated at a number of frequencies resulting in a curve like that in Figure 5.21, known as a signature curve. The signature curve shows the depth of notch required, at a particular frequency relative to the centre of the RF channel, to cause the BER to degrade to 10^{-3}.

Two signature curves are shown in Figure 5.21, one for a 45Mbit/s radio and one for a 19Mbit/s radio, both using the same type of modulation. Note that the notch depth at the centre of the RF channel does not have to be as deep to degrade the 45Mbit/s radio as it does for the 19Mbit/s case. Also, the 45Mbit/s radio is affected by notches as far away as 15MHz while the 19Mbit/s radio is affected over a much narrower frequency range.

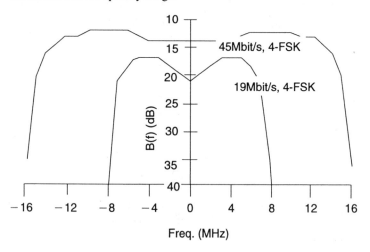

Figure 5.21   Signature curve

A signature curve should be taken for both minimum and non minimum phase fades. Only one is shown for each of the data rates in Figure 5.21 because the minimum and non minimum signatures are nearly identical for the particular system measured.

Originally the signature curve was used to directly calculate the outage time due to selective fading. More recently an intermediate step has been added. First the signature curve is used to calculate the Dispersive Fade Margin (DFM), which is then used to estimate outage time due to selective fading for the particular path under consideration.

The concept of DFM allows one to calculate outage time due to selective fading simply by substituting DFM for Fade Margin in the equation normally used to calculate the outage time of analog radios due to fading. This allowed existing path design programmes to be easily modified for use on digital radio paths. In addition, it allows the susceptibility to selective fading of different radios to be compared on a one number basis instead of an entire signature curve.

Dispersive Fade Margin is calculated using Equations 5.8 and 5.9 from a Bellcore Technical Advisory, where $B_m(f_i)$ and $B_n(f_i)$ represent the minimum and non minimum phase signature curves respectively, and N is the number of f wide segments into which the curve is divided, in order to calculate its area.

$$DMF = 17.6 - 10 \log\left(\frac{S_W}{158.4} \right) \quad dB \tag{5.8}$$

$$S_W = \sum_{i=1}^{N} \left(e^{-B_n(f_i)/3.8} + e^{-B_m(f_i)/3.8} \right) \Delta f \tag{5.19}$$

For example, if the signature curve of Figure 5.21 for the 45Mbit/s radio is broken into 33, 1-MHz wide segments the result is as in expression 5.10

$$\sum_{i=1}^{33} e^{-B_n(f_i)/3.8} = 0.938 \tag{5.10}$$

Since the minimum and non minimum phase signatures are the same, S_W is twice this value, 1.876. Putting this into Equation 5.8 results in DFM = 36.9dB.

5.2.2.3 *Composite fade margin*

The significance of DFM depends on its value relative to the normal system fade margin which is often referred to as Flat Fade Margin (FFM). To see this, DFM and FFM are combined into a single term known as Composite Fade Margin (CFM) using Equation 5.11.

$$CFM = -10 \log \left(10^{-FFM/10} + 10^{-DFM/10} \right) \qquad (5.11)$$

A plot of CFM vs. FFM for DFM = 30 and 40dB is shown in Figure 5.22. Note that when FFM = DFM the Composite Fade Margin is 3dB less than FFM or DFM. Also note that increasing FFM beyond DFM + 6dB results in very little further improvement in CFM.

The Composite Fade Margin may now be substituted for Fade Margin in the equation for Fractional Annual Outage Probability as in Equation 5.12.

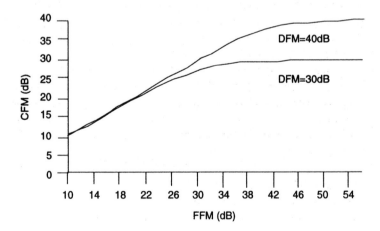

Figure 5.22   CFM vs. FFM

$$P = c \times b \times 2.5 \times 10^{-6} \times f \times D^3 \times 10^{-CFM/10} \qquad (5.12)$$

For example, with CFM = 35dB, average terrain (c=1), average climate conditions (b=0.25), f=18.5GHz, and a path length of 10 miles, the resulting Outage Probability is, P=3.66x10^{-6}. This is equivalent to 115 seconds per year. This outage time is due to the combined effects of Flat and Dispersive fading because we used CFM in the calculation.

5.2.2.4 *Conclusions about DFM*

When designing 18 and 23GHz paths, Dispersive Fading may dominate when using the highest data rate radios (DS3, 45Mbit/s), with RF frequencies in the 18GHz band, in areas of the country that are particularly dry (low rain rates).

At intermediate data rates, 19Mbit/s for example, in the same dry climate at 18GHz, the dispersive outage may be about the same as the rain outage. At the lower data rates (DS1, DS2) rain outages should dominate in all climates.

It also appears that rain outage will still be the largest factor in all 23GHz paths. When Dispersive Fading is dominant, increasing FFM beyond 6dB above DFM will not significantly improve outage time. When Rain Fading is dominant, increasing FFM beyond this point will improve outage time.

In summary, the Dispersive Fade Margin and specification is not as significant at 18 and 23GHz as it is at 2 and 6GHz because a typical path's outage time is still dominated by Rain Fading at the higher frequencies.

5.3 Digital and analogue microwave systems

Though it is in everyday use in the Common Carrier Services, digital microwave and multiplex is beginning to replace analogue microwave and multiplex in the Private Operational Fixed Microwave Services. The private user, however, may question the need to replace his analogue system or to acquire a new digital system. There are system cost differences and system performance differences. This section will address mainly the system performance differences.

Hybrid systems transmitting data on analogue voice channels on analogue microwave are not uncommon. Another type of hybrid system is formed by the use of T-1 baseband modems, which permit the transmission of DS-1 signal over analogue microwave facilities. Certain of these configurations will also be considered.

5.3.1 Noise

The distance over which one hop of microwave radio can provide reliable communications is limited essentially by line-of-sight distance, i.e. the antennas of the two stations must be within visible range of each other. Within this limitation, the distance is further limited by multipath fading due to atmospheric effects which becomes rapidly more severe as the path length increases, and is worse for higher frequencies. For even higher microwave frequencies the phenomenon of rain attenuation becomes more important than multipath fading.

In short, the typical microwave path is relatively short, shorter than line-of-sight, and is designed for the desired path availability (these topics have been analysed in earlier sections). Thus, in order to extend a communications system to reasonable sizes, multiple hops, or relays, of microwave are required.

The concept of cumulative noise in analogue systems is well known. For analogue microwave/multiplex systems, the basic noise level is determined by the Frequency Division Multiplex channel equipment, and the microwave radios over which the channel is conducted adds to this noise. The noise of each microwave hop adds to what is already present, thus, the longer or larger the system, the worse the signal-to-noise ratio becomes. There is no easy way to avoid this phenomenon or to remove the accumulated noise and distortion.

It is commonly assumed that one of the great advantages of digital microwave/multiplex (and digital communications systems, in general) is that, unlike analogue systems, noise does not accumulate.

Within limits, this is true, however, digital microwave/multiplex systems will be shown to have other noise phenomenon to consider.

Noise phenomena may be used to compare analogue and digital microwave systems on a static basis which is the 'normal' operating condition, and also on a dynamic basis where microwave signal fading reduces system availability.

5.3.2 Voice

Either analogue or digital microwave may be used to transmit voice communications effectively. However, the technical jargon used to describe system performance differs somewhat for analogue and digital microwave systems, thus tending to make comparisons based on voice communications difficult.

In this chapter, we will use the concept of Average Voice Power-To-Noise ratio as a common parameter to compare the voice channel performance of analogue and digital equipment.

5.3.3 General comparisons of analogue and digital performance

It is informative to note that in the digital microwave/multiplex system, both voice and data channels are affected similarly by microwave signal fading to the receiver threshold below which performance is unsatisfactory (see Figure 5.23). This is in contrast to the difference in voice channel and data channel performance in an analogue system, where voice channel performance remains usable (above 15dB Average Voice-To-Noise) when the received signal level has faded below the level which will cause data channel outage. Figure 5.24 illustrates this for a 32 QAM 12Kbit/s data channel which occupies a voice channel on an analogue microwave system.

The digital system equivalent to baseband frequency slot allocation is time slot allocation, which makes absolutely no difference in channel performance, unlike the analogue FM system where the highest baseband frequency slot usually has the worst noise. In the analogue system, serious consideration must be given to the assignment of baseband frequencies, especially when assigning such sensi-

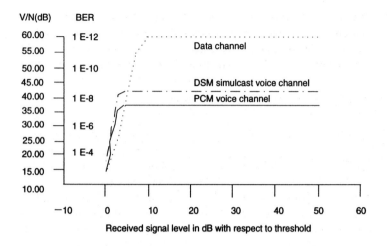

Figure 5.23 Average Voice-To-Noise and data channel BER vs. received level on a digital microwave system

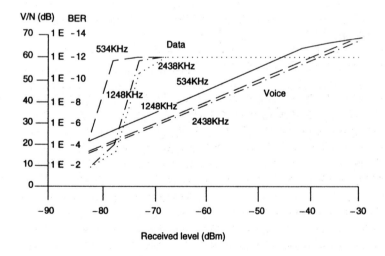

Figure 5.24 Average Voice-To-Noise and 32 QAM data channel BER vs. received level on a 600 channel analogue FM system (with emphasis)

tive devices as the 256 QAM T-1 baseband modem. This type of baseband modem may begin to make bit errors for received levels as much as 20dB above the analogue microwave FM receiver threshold, unless it is assigned to the low end of the baseband spectrum.

5.3.4 System performance vs. size

Further insight may be gained by comparing digital and analogue multi-hop systems. Table 5.1 shows the estimated Average Voice-To-Noise ratio of a voice channel on a hypothetical multi-hop analogue FM microwave system for various numbers of hops and parameters given in Table 5.2. The Average Voice-To-Noise ratio on a digital system, however, is limited by the A/D conversion technique used in the channel equipment; for companded PCM this is about 37dB. In this example, the number of hops where the analogue system Average Voice-To-Noise ratio has diminished to 37dB due to accumulating noise is 13. Where this crossover point occurs is very strongly affected by the NPR per hop assumed, as will be explained below; therefore, the number 13 should only be used as a rough generalisation.

It would appear that system performance of the hypothetical 13 hop analogue and digital systems are quite similar even when one of the hops experiences signal fading. This is because the Average Voice-To-Noise ratio of the analogue system will remain relatively unchanged, until the system noise level is overcome by the noise of the fading hop. This is illustrated in Figure 5.25, for the worst channel of the analogue system. For larger systems, the all digital approach seems to be equal to or superior to the analogue for voice channel usage.

For a voice channel data modem, and using the same 13 hop system comparison, with one hop fading, the analogue microwave system seems to offer better performance than the digital. In part, this is because the analogue microwave hops used in the comparison have about 10dB lower (better) microwave receiver thresholds than those of the digital hops. Also as might be expected, performance of a data modem on the analogue system improves further if lower frequency baseband slots are used (see Figure 5.26).

Table 5.1 Voice channel performance on analogue microwave

N (hops)	S/N (μW)	S/N (Total)	Average voice/noise (dB)
1	66.11	64.62	49.62
3	60.38	59.93	44.93
5	57.72	57.47	42.47
7	55.97	55.80	40.80
9	54.66	54.53	39.53
11	53.61	53.51	38.51
13	52.74	52.66	37.66
15	51.99	51.93	36.93
17	51.34	51.28	36.28
19	50.76	50.71	35.71
21	50.24	50.19	35.19
23	49.77	49.73	34.73
25	49.33	49.29	34.29
27	48.93	48.90	33.90
29	48.56	48.53	33.53
31	48.21	48.18	33.18
33	47.88	47.86	32.86
35	47.58	47.55	32.55
37	47.29	47.27	32.27
39	47.01	46.99	31.99
41	46.75	46.73	31.73

Table 5.2   Parameters for Table 5.1

Parameter	Value
No. of Channels	600.00
NLR	12.78dB
BWR	28.89dB
S/N — NPR	16.11dB
Nmux	18.00dBrncO
NPR1hop	50.00dB
S/N1hop	66.11dB

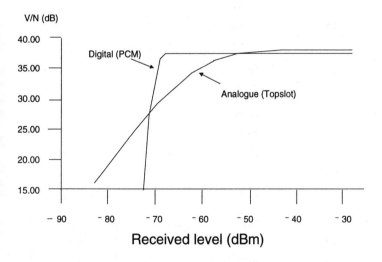

Figure 5.25   Voice-To-Noise performance with one hop fading in a hypothetical 13 hop system. Digital is a PCM channel on DS-3 microwave. Analogue is 2438kHz voice channel of 600 channel FM microwave

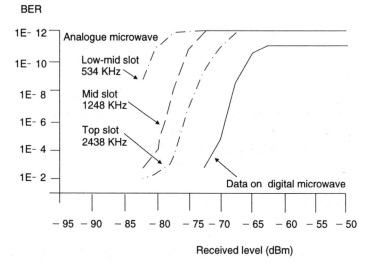

Figure 5.26 Low speed data channel performance on analogue vs. digital microwave. One hop fading in 13 hop system. Digital microwave = DS-3, 10MHz BW. Data channel on analogue = 32 QAM, 12kBit/s. Analogue microwave = 600 channel, FM, 10MHz BW

For high speed data, however, the situation is different. The performance of a 256 QAM T-1 baseband modem is almost always worse than the performance of a T-1 on digital microwave, unless it is located in the low frequency end of the analogue baseband, in which case it may have equivalent performance (see Figure 5.27).

The 13 hop figure used in the above comparison must be accepted only as a typical system size, above which, the digital microwave/multiplex system provides performance advantages, and below which the analogue microwave/multiplex provides performance advantages. As mentioned above, this number is influenced very strongly by certain assumptions regarding the system design, in particular, by the NPR of the analogue microwave hop, as is shown in Figure 5.28. For example, if the NPR per hop is 53dB, the number becomes 26 hops. It also assumes that the majority of the communications traffic is voice.

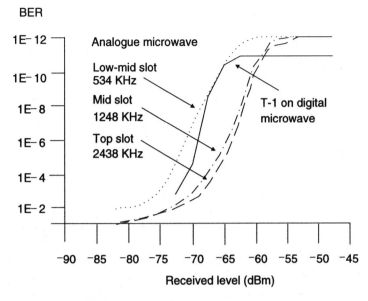

Figure 5.27   High speed data (T-1) performance on analogue vs. digital microwave. 1 hop fading in a 13 hop system. Digital microwave = DS-3, 10MHz BW. T-1 on analogue = 256 QAM modem. Analogue microwave = 600 channel, FM, 10MHz BW

In general, it may be said that there is some system size limit, above which a digital microwave/multiplex system begins to offer significant performance advantages, even for the voice channel users.

5.3.5 Channel capacity comparisons

In the previous examples, DS-3 digital microwave of 10MHz bandwidth was compared with 600 channel FM analogue microwave because the voice channel capacities are similar. However, the data capacities of these types are not similar.

For example, the data channel modems assumed for the analogue examples were 32 QAM, which is used for transmitting 12kbit/s data on a voice channel. Since the microwave system under consideration

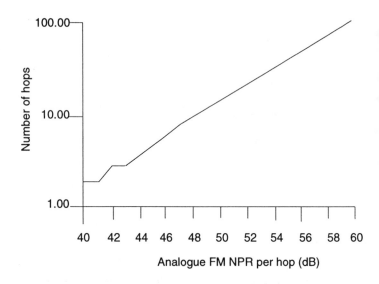

Figure 5.28 An estimate of system size above which TDM/DS-3 digital multiplex/microwave offers potential voice channel performance advantages over analogue FDM/600 channel FM multiplex/microwave

has a voice channel capacity of 600, it might be said to have a data capacity of 12kbit/s × 600 = 7200kbit/s. However, even higher data rate types of data modems may operate at 19.2kbit/s. Therefore, the 600 channel microwave system might also be said to have a maximum data capacity of 19.2kbit/s × 600 = 11520kbit/s. ADS-3 digital microwave system has a capacity of just over 45000kbit/s, or 4 to 6 times the raw bit per second capacity of the analogue FM microwave system. This comparison is shown in Table 5.3.

In this table the digital microwave voice capacity is based on PCM voice channels. The analogue microwave data capacity is based on 19.2kbit/s per voice channel. The analogue microwave T-1 capacity is based on 1 DS-1 per Supergroup.

Another comparison of data capacity might be done on a T-1 basis. In that case, the 600 channel analogue microwave system has a

Table 5.3   Comparison of approximately the same bandwidth and voice capacity analogue and digital microwave

	600 channel FM analogue	*64 QAM/81 QPRS digital*
Bandwidth	10MHz	10MHz
Voice Chan Cap	600	672
Max Data Cap	11.52MBit/s	45.3MBit/s
T-1 Capacity	10DS-1s	28DS-1s
System Gain	110.4dB	103.5 (or 104dB)

capacity of 10 DS-1s, since each DS-1 occupies a minimum of one Supergroup or 60 voice channels. The digital microwave system used in the comparison had a DS-3 capacity, which is the same as 28 T-1 capacity. The digital system thus has 2.8 times the T-1 capacity of the analogue system. This comparison of capacity is also given in Table 5.3.

For this case (similar bandwidth and voice channel capacity), the digital microwave system appears to be vastly superior in data or T-1 capacity over the analogue FM system.

In the preceding examples, the bandwidth and the voice channel capacity of the analogue FM and digital microwave systems were similar. It might also be reasonable to compare analogue FM and digital microwave systems of similar data rate or T-1 capacity. An analogue FM system with a maximum data rate capacity of 11.52Mbit/s or a T-1 capacity of 10 DS-1s, can be compared to a digital microwave system having a capacity of about 12.75Mbit/s and a T-1 capacity of 8 DS-1. (See Table 5.4. The same conditions as Table 5.3 apply.)

In this case the bandwidths are not comparable, the digital microwave system is much more efficient since it requires only 5 or 3.5MHz of microwave bandwidth vs. the analogue FM requirement of 10MHz. However, the digital system has PCM voice channel

Table 5.4   Comparison of approximately the same data capacity analogue and digital microwave

	600 channel FM analogue	*16 QAM/49 QPRS digital*
Bandwidth	10MHz	5(or 3.5) MHz
Voice Chan Cap	600	192
Max Data Cap	11.52MBit/s	12.75Mbit/s
T-1 Capacity	10DS-1s	8DS-1s
System Gain	110.4dB	114.5 (or 113dB)

capacity of only 192 vs. the 600 channel capacity of the analogue system.

For this case (similar data or T-1 capacity) the digital microwave system appears to be superior in terms of bandwidth usage efficiency if primarily data or DS-1s are to be transmitted. However, if primarily voice is to be transmitted, the analogue FM system has a clear capacity advantage.

5.3.6 System gain comparisons

Another significant way to compare analogue microwave and digital microwave systems is in terms of System Gain, i.e. the transmitter power to receiver sensitivity ratio in dB. Higher System Gain is generally viewed as a desirable characteristic of microwave radio equipment, since it is related to the ability of the system to withstand undesirable microwave propagation effects (multipath fading or selective fading, rain attenuation, ducting, etc.). In Tables 5.3 and 5.4 already referred to in the capacity comparison section above, System Gain was also listed for typical microwave radios of the various types.

In the case of the similar bandwidth, similar voice channel capacity comparison (Table 5.3), the System Gain of the analogue FM type is shown to be superior. This is due to the poorer microwave receiver

Table 5.5 Comparison of approximately the same System Gain analogue and digital microwave (for the same conditions as Table 5.3)

	600 channel FM analogue	*16QAM digital*
Bandwidth	10MHz	10MHz
Voice Chan Cap	600	384
Max Data Cap	11.52MBit/s	25Mbit/s
T-1 Capacity	10DS-1s	16DS-1s
System Gain	110.4dB	111.5dB

threshold of the bandwidth efficient digital microwave system. This bandwidth efficiency was necessary for the digital microwave system to achieve a similar voice channel capacity and bandwidth.

In the case of the similar data or T-1 capacity comparison, the System Gain of the digital microwave type is slightly superior. This is because the digital microwave is a more efficient mode for transmitting high speed data, and requires less bandwidth for a similar data rate. Less bandwidth equates to less noise in the passband of the digital microwave receiver, and a relatively good receiver threshold.

Table 5.5 adds a new comparison, where similar System Gain analogue FM and digital microwave are compared. In this case, the digital system is superior in terms of data or T-1 capacity (by up to two times), but the analogue FM system is superior by about the same factor (up to two times) in terms of voice channel capacity. The bandwidths are similar or the same.

5.3.7 Summary

Advantages of digital microwave are:

1. No difference between voice and data threshold to consider.

2.   No baseband slot allocation difference to consider.
3.   Performance advantage for voice users in large systems and for high speed (T-1) data.
4.   Given similar bandwidth and voice channel capacity: much high data or T-1 capacity.
5.   Given similar data or T-1 capacity: much more bandwidth efficient for data or T-1 transmission.
6.   Given similar System Gain and bandwidth: higher data or T-1 capacity.

Advantages of analogue microwave:

1.   In small to medium sized systems: superior voice channel performance.
2.   For most system sizes: performance advantage for voice channel data modem users.
3.   Given similar bandwidth and voice channel capacity: much higher System Gain.
4.   Given similar System Gain and bandwidth: higher voice channel capacity.

Though rare exceptions may be found to these conclusions, the summary of advantages above will be found to be applicable given the bandwidth restrictions and the traffic requirements of the Private Operational Fixed Microwave Services.

5.4 System specification and equipment design

5.4.1 The role of microwave in systems

For the owner/user of a microwave system, performance and reliability are of utmost importance. The usual function of a microwave system is to serve as an interconnecting medium for other equipment which is considered the 'real' communications system. For example, microwave is commonly used to convey multiple telephone conversations from one location to another, possibly via one or more inter-mediate locations or sites. The telephone users do not expect to hear crosstalk from other telephone conversations sharing the same micro-wave system. Other sources of noise or interference due to the microwave system should also be imperceptible. To the user, the fact that microwave is being used is irrelevant, and normal landline telephone quality is the minimum expected.

Another example is where microwave is used to interconnect data devices (computer terminals, keyboards, CRT displays, etc.) with control and monitoring devices at remotely located sites along a pipeline. If the microwave interconnection somehow caused data errors in outbound transmission, the terminal operator's controlling commands could be misinterpreted at the remote site resulting in a wrong and possibly costly erroneous action taking place (such as opening or closing the wrong valve, etc.). Similarly, if the operator received erroneous gauge or meter readings due to a characteristic of the microwave system, he might be encouraged to take an erroneous action.

For the user, the microwave interconnection should be a transpar-ent portion of his total system. This is the objective of the microwave equipment designer and the microwave system designer. In order to achieve system transparency:

1.   The equipment specifications must be consistent with trans-parency in the system.
2.   The equipment must meet or exceed its technical performance specifications.
3.   The microwave system must be designed to make effective use of the equipment's performance.

5.4.2 Important system specifications or characteristics

One of the most important system characteristics affecting both performance and reliability is the area of coverage. This is true because the length and number of the microwave hops in the system has a direct effect on other desired system specifications such as signal-to-noise ratio and distortion in a voice channel, or Bit Error in a data channel.

The length and number of microwave hops has a strong influence on system availability due to probability of system outage from microwave propagation phenomenon.

Also, the area of coverage determines the amount of microwave equipment required which, since no equipment is perfectly reliable, strongly affects the system availability due to outages from equipment failure.

All of the above, i.e., signal-to-noise ratio, distortion, Bit Error Rate, and system availability are components of transparency, which, as previously discussed, is the most desirable overall characteristic of the microwave portion of a system. These components are determined by the microwave equipment specifications, and just as importantly, by microwave system design.

5.4.3 Important equipment specifications

Many microwave equipment specifications are interrelated, as far as their effects on system performance are concerned, and most of them are of importance. The most important of the equipment specifications, if any can be so identified, is System Gain. System Gain's importance is that it is a major controlling factor in system design, system performance, system availability and system cost, i.e. almost everything of interest.

Prior to defining System Gain and delving deeper into the subject, the equipment specification referred to as NPR must be listed as the next most important. Equipment NPR affects all of the system characteristics mentioned for System Gain, with the exception of system availability. Like System Gain, the definition of NPR will be given later and the subject examined further.

Let it be said at this point that improvements in both System Gain and NPR are highly sought by both the system designer and the equipment designer. Thus, the system requirements influence progress in equipment design; yet, the status of the equipment design determines the system characteristics. Improvements in both have been allowed mainly by advances in microwave semiconductor technology.

5.4.4 System gain

The microwave equipment performance parameters of transmitter power output and receiver sensitivity have been combined into a single specification — System Gain. System Gain is the decibel difference between transmitter power and receiver sensitivity. Thus higher transmitter power, improved receiver sensitivity, or both, will give a higher System Gain figure. High System Gain is considered beneficial for the following reasons:

1.　　It may allow longer hops.
2.　　It may allow smaller antennas.
3.　　It may allow improved S/N ratio.
4.　　It may allow less frequent outage due to multipath fading.
5.　　A combination of 1, 2, 3 and 4 may be obtained.

The well known inverse-square relationship between free-space path loss and path length expressed in decibels is as in Equation 5.13 or 5.14.

$$Path\ loss\ (in\ db) = 36.6 + 20 \log d\ (in\ miles)$$
$$+ 20 \log f\ (in\ MHz) \qquad (5.13)$$

$$Path\ loss\ \alpha\ d^2 \qquad (5.14)$$

A microwave receiver requires a minimum received level, thus increased System Gain may be used to allow longer hops. This may seem obvious, but what is not so obvious is that longer hops can result in less microwave equipment required to cover a given distance, and that less equipment results in higher reliability. The inherent and unavoidable failure rates of cascaded equipment adds. The Mean Time Between Failure (MTBF) of the basic multi-hop microwave system is the inverse of this total failure rate as in Equations 5.15 to 5.17 where n is the number of similar microwave radios in series.

$$\lambda_T = \lambda_1 + \lambda_2 + \lambda_3 + \ldots + \lambda_n = n\,\lambda_i \qquad (5.15)$$

$$MTBF_T = \frac{1}{\lambda_T} = \frac{1}{n\,\lambda_i} \qquad (5.16)$$

$$MTBF_T \ \alpha \ \frac{1}{n} \qquad (5.17)$$

If longer hops are not needed or desired, it may be possible to use smaller antennas. In the USA the FCC, however, places a limit on the smallest antenna that can be used. A commonly used formula for dish antenna gain is as in Equation 5.18 or 5.19.

$$\begin{aligned} G \ (in\ dB) = \ &20 \log f \ (in\ MHz) \\ &+ 20 \log D \ (in\ feet) \ - \ 52.6 \end{aligned} \qquad (5.18)$$

$$Antenna\ gain \ \alpha \ D^2 \qquad (5.19)$$

Thus it can be seen that high System Gain may allow the use of the minimum size (diameter) dish antenna. This antenna not only costs less, but decreases stress due to wind loading on the supporting structure. In this manner, System Gain can reduce system cost and increase system reliability.

Given a path length and antenna size, increased System Gain will increase the level of the signal arriving at the microwave receiver. For the FM microwave receiver, this may improve the S/N ratio in a voice channel, or the BER in a data channel, if the intrinsic noise of the radio is not yet the limiting factor. This can be seen from the relationship between received level and S/N for an FM receiver, as in Equation 5.20, where C is the received level. $139 = (-KTB)$ in dBm in a 3kHz bandwidth. F is the receiver noise figure. d_{rms} is the test tone deviation. f_{mod} is the test tone frequency.

$$\begin{aligned} S/N \ (dB) = \ &C \ (dBm) \ + \ 139 \\ &- \ F \ (dB) \ + \ 20 \log \frac{d_{rms}}{f_{mod}} \end{aligned} \qquad (5.20)$$

Improved outage due to multipath fading can be the result of increased received signal level.

It should be apparent that System Gain is an extremely important equipment specification. Yet it is possible to have too much System Gain at some point and actually begin to harm system performance, reliability, or cost. For example, high transmitter power may be neither desirable nor beneficial as shall be shown in the following section.

5.4.5 Transmitter power output

On the subject of transmitter power output, it should first be realised that microwave transmitter power output has a direct and strong impact on all of the parameters of concern, namely equipment cost, system cost, reliability and performance. In addition, though the technology of generating transmitter power has changed and generally improved, the amount of transmitter power has not significantly increased. One interpretation of this is that transmitter power is still very costly, there has been no great need for increased transmitter power, and that system gain improvements, as we shall see later, have been made by relatively inexpensive improvements in receiver sensitivity.

The predicted increase in transmitter power output in the future may occur due to the fact that there is very little further improvement in receiver sensitivity to be had. The exact technology that will be used remains to be determined, but GaAs FET amplifier technology looks most promising. Unfortunately, it seems to be a law of nature that along with the benefits possible from increased transmitter power there are also severe penalties in cost and reliability. The thrust of new technology is generally to reduce those penalties.

Presently, the cost for transmitter power is strongly dependent on microwave frequency, i.e. power costs much more at higher frequencies.

In many systems, present microwave transmitter output power is already more than adequate, and in some cases is sufficient to degrade system performance by severely overloading the microwave receivers. Often the excess transmitter power is thrown away by using power attenuators. In these cases, the potential benefits to be gained by using low power are not realised. Not only is low power significantly less costly, but also the associated decreased equipment temperature, decreased parts count, and decreased d.c. load can yield improved equipment reliability.

For example: a typical solid-state power amplifier (either 2GHz or 6GHz) consists of several stages internally, and may have an MTBF[3] of 100,000 hours. A typical microwave installation might not require the power amplifier at all, but may require transmitter power output of no more than 10 to 100 milliwatts. If the MTBF of the transmitter, including the power amplifier, is known, the MTBF of the transmitter without the power amplifier can be estimated using Equation 5.21 as in Equation 5.22 for $MTBF_{XMTR\ W/PA}$ equal to 26000 hours or approximately three years.

$$MTBF_{XMTR\ W/O\ P.A.} = \cfrac{1}{\cfrac{1}{MTBF_{XMTR\ W/PA}} - \cfrac{1}{MTBF_{PA}}} \qquad (5.21)$$

$$MTBF_{XMTR\ W/O\ P.A.} = \cfrac{1}{\cfrac{1}{26000} - \cfrac{1}{100000}} \qquad (5.22)$$

Thus, a low power transmitter, capable of only 10 milliwatts or so should be preferred over a high power transmitter (1 watt or more) followed by a power attenuator, when the system characteristics (path length, frequency, antenna gain etc.) permit.

5.4.6 Receiver characteristics

The receiver characteristic normally considered is sensitivity, i.e. weak signal performance. System Gain has been improved through the years by improving receiver sensitivity. Unlike transmitter power output, improved receiver sensitivity is relatively cheap. Today's low noise preamplifier transistors are actually cheaper and better performing than the devices of the past (tunnel diode preamplifiers, parametric amplifiers, low noise mixers, etc.). Unlike transmitter power output, there is a law of nature limitation on receiver sensitivity: this is the random noise (KTB noise) that, for a given bandwidth, is only dependent on temperature. It represents the '0dB Noise Figure' limitation, which in turn directly determines the best achievable receiver sensitivity.

Today's receivers are only perhaps 2 to 3dB from this limit in their preamplifiers, and also have further unavoidable limitations due to preselector filtering losses. In recent years, the Noise Figure of typical microwave communications receivers has followed the curve of Figure 5.29. This also shows the curve extended in the future towards the approachable, but unachievable limitations presented by KTB noise and filter losses.

Due to possible future improvements in both transmitter power output and receiver sensitivity, only 5 to 6dB increase in System Gain is anticipated. This improvement in System Gain may be used for worthwhile improvements in system characteristics by any of the means discussed earlier.

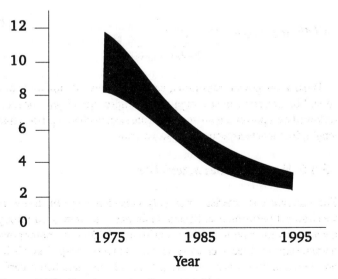

Figure 5.29   Receiver noise figure trends

It should be noted, however, that barring unforeseen technological breakthroughs, it is not and probably will not be wise to design systems for significantly higher received levels than specified for the equipment. This is because, not only do the S/N improvements level out at the intrinsic limitations of the transmitter and receiver, but also NPR (of which S/N and distortion are components) may actually degrade due to receiver overload or saturation. Generally speaking, older receivers are worse in this regard, but receiver overload characteristics are seldom specified even today.

5.4.7 Noise Power Ratio (NPR)

Some references to NPR were made previously; this section will define the term and examine the subject in some detail. NPR is one of the most important microwave equipment specifications. NPR is a major controlling factor in system design, system performance, and system cost. NPR is expressed in decibels and a larger number is better. It is somewhat a function of System Gain in that it is partly determined by the level of the received microwave signal. Improved NPR primarily allows improved S/N and lower distortion in a voice multiplex channel on a microwave system. High NPR is necessary for low Bit Error Rates on data channels. NPR, therefore, is a prime factor in determining the desired transparency of the microwave portion of a system.

NPR stands for Noise Power Ratio. This is indicative of the test procedure performed in making the measurement. The basis for the measurement of NPR is that microwave systems carry many independent communications channels simultaneously. All of these channels not only are affected by the noise floor of the transmitter and receiver, but also create noise-like interference above this level by the creation of harmonic and intermodulation distortion due to phase and amplitude non linearities of the microwave system. That is, harmonics of each channel are produced, and most importantly, each channel intermodulates with every other channel creating sum and difference frequencies.

In the NPR test, the multiple channel baseband signal is represented by a noise generator adjusted for the equivalent power level

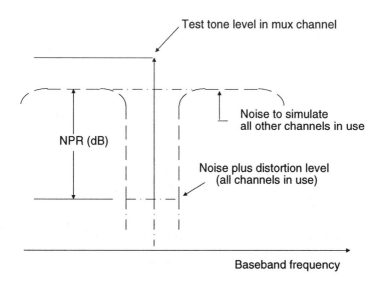

Test tone level in mux channel

NPR (dB)

Noise to simulate
all other channels in use

Noise plus distortion level
(all channels in use)

Baseband frequency

Figure 5.30   NPR test

and frequency spectrum (see Figure 5.30). At the system output the power of this noise is measured in the channel under test. The noise generator spectrum is then notched at the frequency of the channel under test. At the system output, the residual noise power is measured in the channel under test. The ratio of these two readings is the Noise Power Ratio. An ideal, noiseless, distortionless system would have an infinite NPR.

Due to the nature of NPR, all sources of channel degradation are included. The microwave radio specification includes transmitter idle noise, receiver idle noise, receiver thermal noise, transmitter distortion, and receiver distortion. It is measured from baseband input jack to baseband output jack of a transmitter-receiver pair.

Signal-to-noise ratio limited by idle and thermal noise could be improved by increasing the baseband signal level. Normally, this cannot be done without exceeding the authorised bandwidth of the microwave transmitter, unless the entire baseband channel capacity of the microwave system is not going to be used.

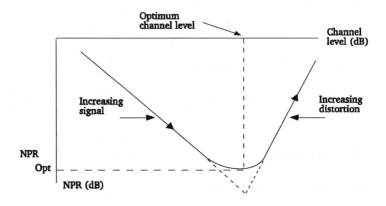

Figure 5.31   The bucket curve

Unfortunately, raising the baseband signal level also increases the distortion level which tends to degrade NPR. Therefore, there is an optimum baseband level which is sufficiently high to gain the maximum improvement over intrinsic and thermal noise without creating an excessive amount of distortion. This situation is represented by the 'Bucket Curve' (see Figure 5.31), and very likely the recommended level given by the equipment manufacturer is at or near the optimum level.

If the received microwave signal level could be increased above 'normal' level perhaps due to increased transmitter power, the receiver thermal noise contribution could be decreased. Equivalently, improved receiver sensitivity, given a fixed received level, would also decrease the receiver thermal noise contribution and result in improved NPR. This improvement could be achieved if NPR is not already significantly limited by idle and intermodulation distortion noise. Thus, it can be seen that System Gain may somewhat control NPR.

It should not be forgotten that in the actual system, there are yet more noise contributors, namely co-channel and adjacent channel interference, and echo distortion. These will not be discussed here, but are mentioned as a reminder that system NPR is partly in the hands of the system designer.

It is interesting to note also that if idle and intermodulation could be entirely eliminated by clever design, and given that baseband levels are fixed at their current levels by bandwidth considerations, and given that received levels remain essentially at the current levels due to receiver overload considerations, the typical NPR could improve several decibels.

5.5 References

Berman, M.J. (1993) Line of sight communications for business, *Telecommunications*, February.

Bierman, H. (1990) Non-military microwave applications, *Microwave Journal*, April.

Brett, M. (1994) Line-of-sight surveys for microwave radio relay links, *Mobile Europe*, March.

Colombo, G. and Podolak, T. (1994) Millimetric wave transmission, *Electrical Communication*, 4th Quarter.

Edwards, T. (1990) Digital microwave transmission systems, *Telecommunications*, December.

Edwards, T. (1991) Terrestrial microwave link systems, *Communications International*, August.

Hubble, R. (1995) A system for all seasons, *Mobile Europe*, April.

Nannicini, M. et al. (1994) SDH microwave transmission — network application, key technologies and products, *Electrical Communication*, 4th Quarter.

Sohlo, M. (1993) Microwave on the hop, *Communications International*, September.

6. Point to multipoint urban and rural radio

6.1 Introduction

Radio systems have the following advantages over other transmission media such as coaxial and optical cables:

1.   Radio systems can be installed relatively quickly for permanent or temporary applications.
2.   Equipment can be recovered and re-used elsewhere.
3.   Ideal for use in hostile environments, e.g. marshes, hilly terrain, over water paths etc.
4.   Radio is more attractive to use when the demand for a new service is not certain or difficult to forecast.
5.   Radio can provide media diversity to cable systems on routes requiring highly reliable communication.

The disadvantages of radio are as follows:

1.   Generally line-of-sight is required between terminals.
2.   Radio signals distort and/or attenuate due to the weather.

Point-to-point radio systems are suitable for relatively large transmission capacity. However, for kilobit rate services it is difficult to provide spectrally efficient point-to-point systems at microwave frequencies for two reasons: it is difficult to make low loss filters with extremely narrow bandwidths, and power source instabilities could result in a relatively significant transmission bandwidth requirement. These disadvantages are overcome by using point-to-multipoint systems. There are two types of multipoint systems, one designed for urban use and the other for rural applications.

6.2 Urban multipoint systems

For urban applications such as data and integrated services digital network (ISDN) access, multipoint systems are usually required to operate over relatively short hops of up to 7km to 10km. This requirement, coupled with the general non-availability of spectrum below about 10GHz (this spectrum is often used for high capacity trunk radio systems) and the desirability to use small, compact, lightweight and unobtrusive roof top mounted equipment has led to the development of a number of multipoint systems in the 10.5GHz to 26GHz bands (Mohamed, 1985; Nakayama, 1985; Omtveit, 1989; Shindo, 1983; and others).

The higher frequencies permit the use of small compact antennas with high gains, and production of highly integrated low cost transceivers. To produce systems with a high spectral efficiency it is essential to use extremely stable but costly oscillators. Thus cost considerations rather than spectral efficiency appear to be the important influencing factors in the system design. Rainfall attenuation becomes significant above about 10GHz and therefore the range and non-availability of these systems are affected by this factor.

6.2.1 System operation

The system comprises a central station (CS) and a number of subscriber outstations (OS) which are usually roof mounted and have line of sight (LOS) paths to the CS. A block diagram of an urban multipoint system is shown in Figure 6.1. The CS would normally be located on a tall building with cable connections to a nearby exchange or central office.

At the CS, low bit rate circuits are time division multiplexed (TDM) into a relatively high aggregate data stream which is then transmitted as a modulated signal at one frequency F_1 in a continuous mode from a wide beamwidth sector antenna. Each roof top radio terminal (or outstation), equipped with a directional paraboloid or Cassegrain antenna, receives this signal and extracts from it the data addressed to it. Return transmissions from all identical OSs occur at a second frequency F_2. Low bit rate customer data are formatted at

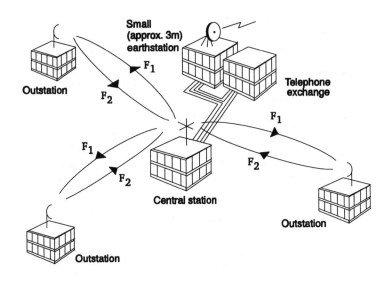

Figure 6.1   Urban multipoint system operation

the same aggregate data rate of the CS for transmission at regular intervals as interleaved bursts with adequate guard time between adjacent bursts. Thus the OSs access the CS in a time division multiple access (TDMA) mode.

The down-link (CS to OSs) modulated signal carries data continuously at the system aggregate rate. In the up-link direction (OSs to CS) burst transmissions are repeated periodically once every 'time frame' period.

Figure 6.2 shows an example of the up-link and down-link time frames for a system designed to operate at 19GHz (Ballance, 1984). Each time slot includes a system overhead such as carrier recovery, symbol timing recovery, signalling, error checking and unique byte identification. This is essential since all the OSs have sources operating at nominally the same frequency F_2. Since the overhead bits for each time slot remains constant, the frame efficiency (i.e. the ratio of total time for subscriber data to the frame time) can be improved by increasing the ratio of useful data to the overhead data. This can be

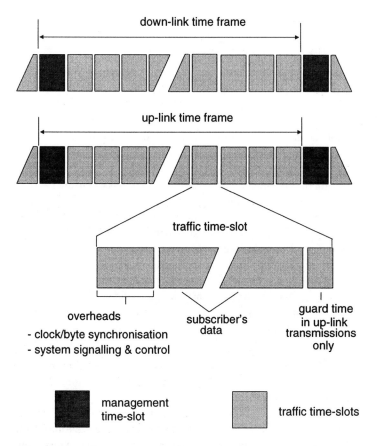

Figure 6.2   Typical frame periods and traffic time slots for TDM/TDMA systems

achieved by increasing the frame period in the up-link direction. A very long time frame period can result in excessive transport delay which may be unacceptable to meet network requirements for speech communication. The choice of time frame period is a compromise between the frame efficiency, system capacity and ease of system implementation.

6.2.2 Channel access

A management or control time slot in the up-link and down-link frame periods (named time slot zero, for instance, in Ballance, 1984) is generally used for tasks such as time slot request/allocation, ranging, error messages and system updates.

The number of time slots determines the number of connections possible at any one time. For private wire or leased circuits, the time slot allocation is fixed or pre-assigned.

On the other hand, time slots may be assigned as and when they are required on demand. In this case two methods of channel access are generally in use, polling and slotted ALOHA random request. In the polling method, the CS continuously polls each OS to determine if a time slot allocation is required. Where a large number of OSs are involved, such a technique requires an unacceptably long time for polling.

In ALOHA, time slot allocation requests are again made during the control time slot period (which may be sub-divided into a number of signalling channels). If the signalling bursts from two or more OSs collide, no acknowledgement is given in the down-link path. Requests are therefore repeated after a random delay duration in order to decrease the probability of further collisions.

6.2.3 Ranging

The burst transmissions from all the OSs, located at various distances, must arrive at the CS as non-overlapping bursts. A small guard time is generally allowed between adjacent bursts in order to cater for slight timing differences occurring in the OSs. When an OS is first installed, it acquires a timing reference from the CS and subsequently determines the time by which its transmission must be adjusted. In one system (Ballance, 1984), when an OS is first installed, it acquires a timing reference signal from the CS, and when commanded by it, the OS transmits a burst in the up-link control time slot. The CS determines its range and instructs the OS to retard its transmissions suitably to correspond to that of an OS located at a maximum distance from it.

6.2.4 System details

A simplified block diagram of an urban multipoint system is shown
in Figure 6.3. The RF equipment at both the CS and OSs is generally
located outdoors with power and baseband (or intermediate frequency, IF) signals carried between it and the indoor baseband equipment.

A number of multipoint systems have been designed at frequencies
between 10.5GHz and 26GHz employing binary or 4-level modulation schemes (Table 6.1). The system aggregate rates vary between
2Mbit/s and 16Mbit/s. These systems are usually used over relatively
short hops (up to 7km to 10km), and therefore the dominant impair-
ment mechanism is rainfall attenuation rather than atmospheric
multipath fading.

Link budgets for such systems are illustrated in Table 6.2 with an
example (Scott, 1984). The receiver sensitivity (–85dBm) is for a

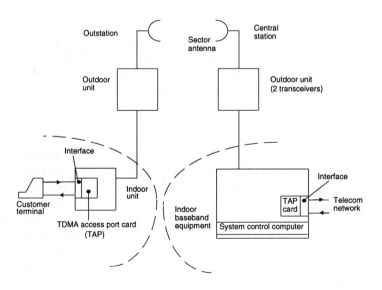

Figure 6.3   A simplified block diagram of a multipoint system

Table 6.1   Urban point-to-multipoint radio systems

| | System | | | |
	10.5GHz	19GHz	23GHz	26GHz
Aggregate bit rate (Mbits/s)	2.048	8.192	0.832	16.384
Modulation	QPSK	2 FSK	2 ASK	FSK (CS to OS); DFSK (OS to CS)
Frame period (ms)	5	7	0.125 (CS to OS) 5 (OS to CS)	1

carrier to noise (C/N) ratio of 11dB since, for binary FSK, the total noise power is the sum of the receiver noise figure (8dB) and thermal noise of −104dBm in a noise bandwidth of 10MHz.

The system fade margin determines the outage due to rainfall. For instance, in the U.K., for a maximum non-availability of 0.01% of an average year, fade margins of 14.2dB and 17.0dB are respectively required at 17.7GHz and 19.7GHz (CCIR, 1990).

The path loss is the sum of the free space loss given by Equation 6.1 and the loss due to atmospheric water vapour absorption (CCIR, 1990a).

$$Path\ loss\ =\ 20\ \log_{10}\left(\frac{4\ \pi\ \ hop\ length}{free\ space\ wavelength}\right) \tag{6.1}$$

6.2.5   Modulation/demodulation methods

Spectral efficiency, ease of implementation and resistance to interference are some of the considerations which can influence the choice

Table 6.2   Link budget for a multipoint system (Scott, 1984)

	Direction of transmission	
	OSs to CS	CS to OSs
Frequency (GHz)	17.7	19.7
(a) Transmitted power (dBm)	21	27
(b) Transmit antenna gain (dBi)	17.5	35.5
(c) Receive antenna gain (dBi)	34.5	18.5
(d) Fixed losses: feeder; filters; radome etc. (dB)	–2	–2
(e) Receiver sensitivity (dBm) for 1 in 1000 bit error ratio	–85	–82
(f) System gain in dB (a)+(b)+(c)+(d)-(e)	156	161
(g) Path loss in dB for 10km	137.9	139.3
(h) System fade margin in dB (f)-(g)	18.1	21.7

of modulation methods. At frequencies above about 10GHz microwave source stability consideration is important in the selection of a suitable modulation method. For instance, for dielectric resonator stabilised microwave oscillators, stabilities of 10^{-4} (temperature range of 5°C to 60°C) around 26GHz have been reported (Ogawa, 1984a). In one system employing binary frequency shift keying (FSK) modulation, the resulting d.c. offset at the CS demodulator has been used to convey a measure of the frequency offset information back to the OS so that the transmit oscillator frequency can be corrected, thus providing a closed loop for fine frequency tuning (Hewitt, 1984; Mohamed, 1985).

Burst transmissions arriving at a CS from all the OSs are nominally at the same frequency. At the CS the carrier phase has to be recovered from the preamble associated with each burst. However, differential demodulation does not require carrier recovery and although it requires a higher C/N ratio, its implementation in the up-link direction is often preferred. The source stability considerations have led to earlier implementations of binary (ASK) amplitude shift keying (Shindo, 1981; Shindo, 1983a; Marchand, 1986). Transmitter power control have also been used to minimise received power variation. However, measurements have shown that ASK requires about 7dB more C/N ratio than for differential frequency shift keyed (DFSK) modulation.

6.2.6 Channel plans

Channel plans are generally different for various systems designed over a frequency range of 10.5GHz to 26GHz (CCIR, 1990b). These are influenced to some extent by the modulation method selected, source frequency variations and the general desirability of avoiding the use of RF filters with extremely narrow fractional bandwidths.

6.2.7 Frequency re-use

The CS antenna is usually a sector antenna with 90 degrees, 120 degrees or 180 degrees coverage. The antenna patterns are generally shaped to ensure that adequate signal is available at OSs which are located at heights much lower (or higher) than the CS (Scott, 1984; Murakami, 1983). In order to increase coverage, other systems at different frequencies may be employed.

Figure 6.4 shows as an example a frequency re-use pattern using 8 frequencies, 90 degrees sector antennas and 4 cell clusters (Shindo, 1981; Mohamed, 1985). In practice, the locations of the systems will inevitably depend on the locations of population centres. Furthermore, the terrain features can have a significant effect on frequency re-use distance. The re-use plan should therefore be used as a guide in determining the frequencies of systems which are widely separated geographically when there is a possibility of using similar systems subsequently in the intervening terrain.

Further variations of the frequency plan are possible by considering the use of both horizontal and vertical polarisations in different

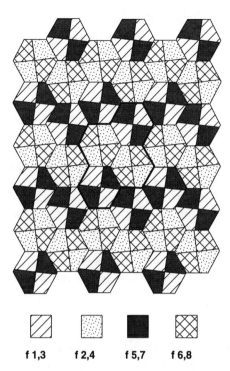

f 1,3   f 2,4   f 5,7   f 6,8

Figure 6.4 One example of a frequency re-use pattern using 90 degree sector antennas

sectors/cells (Murakami, 1983). Interference calculations must allow for co-channel interference from geographically distant systems under differential rainfall fading conditions.

An important consideration in the use of multipoint systems in built up areas is often the non-availability of LOS paths between a subscriber building and a CS. Studies have therefore been carried out in estimating the number of CSs necessary to provide service to a certain percentage of customers in a large city (Ogawa, 1984b). These rely on the availability of actual building density and height distribution information from which the probability of obtaining LOS paths is calculated statistically.

The effect of building reflections has also been examined but it has been shown that a very small percentage of subscribers are likely to be affected from these (Ogawa, 1984b; Rhese, 1986). Other studies (Yoshida, 1984; Dupuis, 1983) have estimated the coverage area for high frequency systems requiring LOS paths between transceivers e.g. as a function of antenna heights.

6.2.8 Applications

The urban multipoint systems are generally used for serving the needs of business customers (e.g. financial and banking community) over relatively short distances in built up cities (Rhese, 1986). Applications include low bit rate data transmission and facsimile. Other applications include access to ISDN and videoconferencing at various bit rates.

The use of high capacity radio systems has been studied for emergency use in situations where, for instance, an exchange serving several thousand subscribers is destroyed by fire or other disasters (Nakayama, 1985).

6.3 Rural multipoint systems

For rural applications, telephony is the main requirement and systems are generally required to operate over long distances, often over hostile, mountainous or marshy terrain. Consequently several systems have been designed to operate in the 1.5GHz to 2.5GHz bands. More recently, following the FCC's authorisation in the U.S.A. of the use of VHF and UHF bands (150MHz, 450MHz and 800MHz) for basic exchange telecommunications radio service (BETRS), 'basic exchange radio' (BEXR) systems (Bellcore, 1989) have been designed for 450MHz band (McGuire, 1989; Mullen, 1989; Lin, 1990). For rural systems hop lengths can be up to 30km to 50km depending on the terrain and the frequency of operation. However, by using synchronised repeaters, the range can be increased further to more than 300km.

Designs of these systems have been influenced by the following factors:

1.   Systems are required to operate over hostile terrain with poor road access.
2.   Inadequacy of primary power supplies makes solar powered system operation highly desirable.
3.   Equipment maintenance must be simplified.
4.   Equipment must have high reliability.
5.   Equipment cost must be low.

6.3.1 System operation

The operation of a typical rural TDM/TDMA system is illustrated in Figure 6.5. Its operation is similar to that of an urban multipoint system. The system comprises a central station (CS) with a wide beamwidth antenna (e.g. omni-directional antenna), and a number of subscriber outstations (OSs) which have directional horn or Yagi antennas.

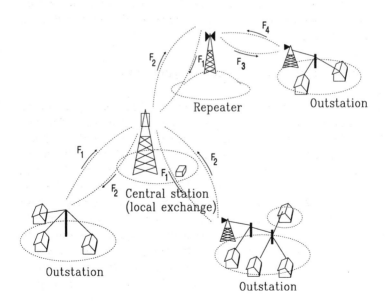

Figure 6.5   Rural multipoint system operation (James, 1990)

The CS is generally located at or near a telephone exchange. The OSs, with line of sight (LOS) paths to the CS, are usually pole mounted and serve a number of subscribers over a distance of a few kilometres via overhead or buried cable pairs.

For telephony, the system effectively provides a transparent path between the telephone exchange and a telephone instrument at a subscriber's premises. Typically the telephony signals from the exchange are encoded in PCM format (A or μ law) by the CS, and these are then time division multiplexed (TDM) before being transmitted in the down-link direction as a modulated signal at one frequency F_1 in a broadcast mode. From this received signal, each OS extracts a timing reference and the data intended for telephony subscribers served by that OS.

In the up-link direction, the telephony signals at each OS are similarly encoded in PCM format, and subsequently suitably formatted for short periodic burst transmission at a second frequency F_2. Thus the OSs access the CS in time division multiple access (TDMA) mode. Although most systems operate in a TDM/TDMA mode, the use of TDMA technique in both directions has also been used (Bonnerot, 1987). This allows the system to operate in a power saving mode at the expense of introducing further time delay.

For most rural applications, the systems are generally designed to operate over considerably longer distances by employing one or more synchronised repeaters in cascade as illustrated in Figure 6.5. The radio hop between a repeater and an outstation (or repeater) requires a further pair of frequencies in order to minimise co-channel interference.

6.3.2 Additional features

The rural multipoint systems generally provide transparent 2-wire analogue telephony connections between telephone instruments and an exchange, although traffic is usually concentrated and de-concentrated digitally over the radio path. The two wire connection gives maximum flexibility but is expensive in terms of line card costs. Therefore consideration has been given to provision of 1.5Mbit/s or 2Mbit/s interfaces from the CS to the exchange (Chas, 1986). At the

CS end 2Mbit/s links between radio and telephony sections have also been used to provide flexibility in siting the radio section remote from the telephony section (De Couesnongle, 1987).

The system design generally permits extension of 2-wire analogue lines a few kilometres from an OS. Some of the systems have additional facilities such as intra-call and automatic ring back mode (Beaupre, 1984). In systems with intra-call facility there is intelligence in an OS to recognise when an outgoing and the corresponding incoming call involve two subscribers who are served by the same OS.

Once the call is established via the exchange, the signalling/metering connection to it via the CS is retained over the management or control time slot, but the OS connects the two subscribers directly, thereby freeing two trunks or time slots for other traffic. This facility increases the traffic carrying capacity.

It is also possible to have intra-call facility even when the link to the CS is temporarily unavailable. Information on local calls during this period may be stored for a limited time and transferred to the CS when the link is restored. In systems with automatic ring back facility, a subscriber attempting to make a call when all the circuits on the radio system are in use, is notified of this fact. Subsequently, the subscriber is offered the opportunity to make a call as soon as a free circuit is available, thereby reducing outgoing lost calls.

Some of the systems designed for rural and urban applications also allow access to ISDN (Bonnerot, 1987; Le-Ngoc, 1989) at 144kbit/s (2B+D). These connections can be at U or S interface in the CS and OSs. Within the radio system, the entire capacity of 144kbit/s per call is not necessarily allocated all the time to the subscribers. Instead the B and D channels may be processed separately so as to use the radio capacity more efficiently, providing B channels as and when needed. Naturally the whole process is transparent to the subscriber.

6.3.3 Channel access and ranging

In urban multipoint systems the path distances are relatively short so that the free space round trip propagation delay (3.3μsec/km) is short enough to enable ranging to be performed automatically by instructing an OS to transmit within the up-link control time slot, for instance, and estimating its distance from the CS.

For rural multipoint systems, the use of synchronised repeaters enables one to cover much longer distances which results in a large round trip propagation delay. Consequently in some systems it is necessary to know the locations of OSs within a few kilometres range from the CS so that the OSs can be ranged automatically (Berndt, 1986; Hart, 1988). In some systems multi-carrier narrowband TDMA systems have been used (Saunders, 1989; Mullen, 1989) so that channel access, still via a control channel, uses a combination of frequency division multiple access (FDMA) and TDMA techniques.

6.3.4 Traffic capacity/grade of service

In a demand assigned system some of the calls will be lost or delayed. The probability of this occurring represents the grade of service (GoS) of the system. The GoS depends on the number of available timeslots or trunks and the offered traffic in Erlangs (E) per subscriber.

Traffic intensity is a dimensionless quantity expressed in Erlangs. If a timeslot is occupied all the time its maximum capacity is 1E. The loss probability, p, of a system is given by the Erlang loss formula, Equation 6.2, where y is the offered traffic in Erlangs and n is the total number of circuits.

$$p = \frac{\dfrac{y^n}{n!}}{1 + \dfrac{y}{1} + \dfrac{y^2}{2!} + ... + \dfrac{y^n}{n!}} \tag{6.2}$$

This is tabulated, for instance, in CCITT, 1989a. Further information on loss probability for network connections is given in CCITT (1989a).

As an example, for a typical busy hour offered traffic per subscriber of 0.07E, the GoS for a system with 15 trunks and 116 subscribers from Equation 6.1 is equal to 0.01.

6.3.5 Channel plans

Most of the rural multipoint systems above 1GHz are designed to operate in the frequency band 1427MHz to 1525MHz, and use an inter-leaved frequency plan. The channel spacing between adjacent channels on the same polarisation depends on the channel bit rate and the modulation method used. CCIR, (1990) gives channel arrangements for systems operating in the 2.3GHz to 2.5GHz band. For rural multi-carrier FDMA/TDMA BEXR systems in the 150MHz and 450MHz bands, channel spacings of 30kHz and 25kHz respectively have been specified (Bellcore, 1989).

6.3.6 System details

Current systems use either binary or four level modulation, and system aggregate rates vary from just below 1Mbit/s to below 5Mbit/s. The number of trunks or time slots varies between 10 and 60. The smaller systems tend to employ a polling technique for accessing time slots, whereas random request slotted ALOHA is used for larger systems (Le-Ngoc, 1985).

The systems are designed for operation in remote locations and consequently low power consumption through use of VLSI and high system reliability are their main features. The outstations can be generally mains powered with battery back up or solar powered.

Two wire telephony, and low bit rate data interfaces are generally provided at the CS and the OSs. At the CS critical circuits such as transceivers are duplicated for increased reliability.

A typical link budget for such systems is given in Table 6.3. At the relatively low operating frequency, rainfall attenuation is less significant. However, the hop lengths can be long (up to about 50km) and therefore atmospheric multipath propagation becomes significant. Since these systems have relatively low bandwidths the narrow bandwidth fading model (CCIR, 1990a) has generally been used to calculate the worst month outage.

The system fade margin, M, is used to determine the non-availability of the system due to propagation effects. The probability (P)

Table 6.3 Link budget for a 1.5GHz multipoint system

Item	Value
Transmission direction	CS to OSs
Frequency (GHz)	1.5
(a) Transmitter power (dBm)	30
(b) Transmitter antenna gain: omni-directional (dBi)	10
(c) Receiver antenna gain yagi (dBi)	16
(d) Fixed losses in dB (transmit/receive end feeders filters etc.)	−5
(e) Receiver sensitivity for 1 in 1000 bit error ratio (dBm)	−94
(f) System gain dB (a)+(b)+(c)+(d)-(e)	145
(g) Path loss in dB for 24km hop	123.5
(h) System fade margin M dB (f)-(g)	21.5

that fading exceeds a given margin, M, is given by Equation 6.3, where D is path length (km), and F is frequency (GHz).

$$P = K\, Q\, F^B\, D^C\, 10 \exp\left(-\frac{M}{10} \right) \tag{6.3}$$

As an example, the parameters K (climatic factor), Q, B and C for North-West Europe are given by $K = 1.4 \times 10^{-8}$, $Q = 1$, $B = 1$ and $C = 3.5$. For this example, $P = 1.4 \times 10^{-8} \times 1.5 \times 24^{3.5} \times 10^{-21.5/10}$, or the availability is given by $[(1-P) \times 100]$ or 99.999% for a path of 24km.

The error performance and availability objectives for digital radio systems used in the local grade portion (between the subscriber and the local exchange) of an ISDN connection is given in ITU-T Recommendation G.821 (CCIR, 1990b). For a 64kbit/s circuit, the bit error ratio (BER) should not exceed 1×10^{-6} during more than 1.5% of any month, and 1×10^{-3} during 0.015% of any month. Furthermore, the total errored seconds should not exceed 1.2% of any month.

The unavailability allowance is generally divided between equipment failure and propagation effects (rainfall for systems above about 10GHz and multipath for lower frequencies). The relationship between errored seconds objectives at 64kbit/s and at the system aggregate rate is given in CCIR (1987).

6.3.7 Modulation/demodulation methods

Although pulse width modulation of speech channel followed by frequency modulation has been used in an earlier system, most systems employ binary frequency shift keying, binary phase shift keying or quarternary phase shift keying methods. At the CS, the phase of the carrier from each OS burst has to be established rapidly. To facilitate this, the OS burst has a preamble for carrier recovery. However, when differential demodulation is used, it is not necessary to recover the carrier. This consideration has led to the use in some systems of differential demodulation at the CS, while coherent demodulation is generally used at the OSs.

Recent BEXR systems have used 16 level differential phase shift keying modulation (Saunders, 1988; Mullen, 1989). Others designed for similar applications have also used 4-, 8-, and 32-level quadrature amplitude modulation in narrower channels of 25kHz (Hampton, 1989; McGuire, 1989).

6.3.8 Power

In a rural environment lack of primary source of power is quite a dominant feature. When power is available it may be subject to interruptions. Therefore subscriber radio systems are designed with the aim of minimising power consumption so that, for instance, solar power operation is possible. The radio equipment generally requires either a.c. power or solar cell arrays with a nominal 12V (or 48V) d.c., the operational voltages being derived by d.c./d.c. converters. Power consumption has been minimised by using, for instance, efficient d.c./d.c. converters, FET amplifiers, low power consuming CMOS arrays and surface mount components (Bonnerot, 1987). In some systems, burst mode transmission is also used in the down-link direction (depending on the frame activity) in order to save power.

6.3.9 Frequency re-use

Frequency planning is a complex exercise (Lawson, 1986). In considering the use of multipoint systems, it is essential to take account of the following:

1. Interference to and from existing systems in the frequency band.
2. A cellular re-use plan must allow for future system growth.
3. The use of repeaters must be considered.

The available channel frequencies may be allocated to form a cellular pattern of 3,4,7,9,12 ... cells (Cox, 1982), with perhaps a smaller number of channels forming an overlap pattern for extra capacity. In addition a frequency re-use pattern may also be established for repeaters.

The frequency re-use pattern is generally based on geometric pattern. However, in practice it is essential to evaluate worst case co-channel interference from cells using similar frequencies in various clusters. Examples of frequency re-use patterns are given in CCIR, 1990b.

6.3.10 Applications

The primary application of these systems is in providing telephony service, often in isolated mountainous villages, deserts or off-shore islands for the very first time. Therefore they are designed to operate in harsh environments over a wide range of temperatures (e.g. -25°C to 55°C). They have found applications in desert areas to provide connections to remote drilling sites with transportable solar powered OSs.

Transportable OSs have also found applications in providing temporary telephony, facsimile and telex services at important events and shows (Holm, 1988). Other applications include emergency service restoration, provision of SCADA (supervisory control and data acquisition), and data services in rural areas.

The more recently developed systems offer facilities for connecting subscribers to ISDN. With the higher capacity systems, the OSs are capable of serving a large (> 100) number of subscribers. Therefore, it is possible to consider the advantages of installing a multipoint OS rather than a small exchange (Blake, 1989). Combined with an intra-call option, the OS then behaves effectively as a small exchange.

6.4 Economics of urban and rural systems

Generally, urban systems can potentially be high revenue earning systems since they would be used for serving the needs of business subscribers. On the other hand, rural systems give a poor return on investment, and the decision to employ them is often a political one.

The cost per circuit depends on the system configuration which, in turn, depends on subscriber distribution and terrain features requiring possible use of repeaters to overcome LOS path problems. In general, the initial cost of the system is significant when only a few circuits are provided and the cost of the CS predominates. Subsequently, however, the cost per circuit decreases as more circuits are added in a flexible manner as the demand grows. Generally, the cost per circuit is low when a large number of circuits are provided from an outstation, since the cost of the common hardware (e.g. RF transceiver) is shared (Figure 6.6).

In general, the cost of a circuit provided by radio is constant as a function of distance. On the other hand, cable costs generally increase with distance. However, cost comparison with alternative media such as fibre or coaxial cable depends on the extent of existing telecommunications infrastructure and the operational environment (CCITT/CCIR, 1976).

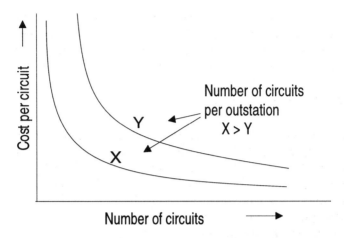

Figure 6.6   Cost per circuit as a function of number of circuits

6.4.1 Transport delay considerations

Point-to-multipoint systems introduce group delay due to the packetised nature of the system. Thus the round trip delay of a system is the sum of the up-link and down-link frame periods plus additional processing delay which could be 10% to 30% of the frame delays. In addition, where synchronised repeaters are used to increase the system range, the propagation delay of about 3.3 microseconds per kilometre also becomes significant since the maximum delay corresponds to that of an outstation located at a maximum distance from the CS. A combination of excessive delay and mismatches from a far end 2 to 4 wire hybrid can give rise to echo which can hinder communication unless echo control devices are introduced in the network.

CCITT (1989b) give guidance on the use of echo control devices for telephony when the one way group delay exceeds 25 milliseconds. In addition to this, in a national network, consideration has to be given to the use of echo control devices at gateway international switching centres, and the possible consequences of using echo control devices in cascade. In a national network, where two or more

Public Telecommunications Operators provide service, the delay limits may preclude the use of the systems which introduce large group delays, unless suitable echo control devices are permitted without affecting the telephony performance significantly.

6.4.2 Future trends

In future systems, three issues are likely to be important: cost, spectrum efficiency and delay associated with multipoint systems. The use of higher level modulation schemes (e.g. 16 QAM) would lead to an increase in spectral efficiency and potential reduction in delay. However, counter measure techniques now commonly used on trunk radio systems would then need to be carefully adapted to increase the system range and reduce costs.

TDM/TDMA techniques are also being employed in fibre systems for application in the local loop (Oakley, 1988; Hoppitt, 1989). In such systems, the use of passive splitters ensures that no electronics is used between the exchange and the subscribers' premises. There are advantages in making the TDM/TDMA sub-systems for providing telephony over passive optical networks (TPON) such that they can be readily employed for radio use. This provides additional flexibility in employing the most appropriate medium (radio or fibre) in different circumstances whilst using common TDM/TDMA subsystems and network interfaces.

Cellular radio systems at 450MHz and 900MHz have proved to be popular in Europe. Although such systems do not provide speech quality as good as that over fixed radio or cable, their popularity is likely to grow rapidly since they offer additional roaming facility. Their use in fixed network (Stern, 1987) may be hindered by lack of spectrum, regulatory constraints and higher infrastructure costs resulting in higher charges than for fixed service.

In Europe the establishment of personal communications networks (PCN) over the next few years in the 1.7GHz to 1.9GHz band may well pose a challenge to multipoint radio systems providing fixed services. The envisaged mass markets for these networks will inevitably lead to hardware price reductions and reflect in lower telecommunications costs to the subscribers. Manufacturers of multipoint systems may adapt the cheaper hardware, for instance, to provide complete exchange to subscriber premises radio solutions.

To date, most multipoint radio systems have aggregate rates of less than 20Mbit/s. The synchronous digital hierarchy (SDH) multiplexer offers flexible 1.5Mbit/s or 2Mbit/s drop and insert facility together with maintenance/management protocols at 155.22Mbit/s (STM-1) bit stream. Therefore designs of multipoint radio systems with STM-1 aggregate data rate may well become attractive in future (Richman, 1990).

The role of satellites in rural communications is mentioned in passing. Interest in this has increased with the recent announcement of plans for Iridium, a digital voice/data system in the 1GHz to 2GHz range providing global coverage via 77 low orbiting, networked satellites (Satellite Communications, 1990). In such a system a LOS path would be possible to every point on the earth's surface. However, it will be late nineties before service commences from such a system.

The search for ways of decreasing radio system power consumption, increasing its reliability and reducing its costs (such as with the widespread use of monolithic microwave integrated circuits) will no doubt continue.

6.5 References

Bailey, V.L. (1986) A multipoint radio for local networks, *European Conference on Radio Relay Systems, ECRR*, Munich, West Germany, pp. 276-283.

Bailey, V.L. and Ogborne, J. (1987) A multipoint radio for local networks, *IEEE International Conference on Communications, ICC'87*, Seattle, U.S.A., pp. 332-335.

Ballance, J.W. (1984) A low cost TDM/TDMA subsystem for point-to- multipoint radio applications, *Br Telecom Technol J.*, **2**, (2), pp. 19–24.

Bannister, G. and Capewell, C. (1988) The Australian digital radio concentrator system — DRCS *IEE International Conference on Rural Telecommunications*, London, U.K., Conference Publication No.286, pp. 56-62.

Battistig, G. et al. (1985) A simple point-to-multipoint subscriber radio system in the 1.5GHz band, *IEEE International Conference on Communications, ICC'85*, Chicago, U.S.A., pp. 23.3.

Battistig, G. Marczy, A. and Rona, P. (1986) IER, a subscriber radio system in the 1.5GHz band, *Proceedings of the Eighth Colloquium on Microwave Communications*, Budapest, Hungary, pp. 19-20.

Beaupre, D.M. (1983) Five years of TDMA subscriber radio *IEEE International Conference on Communications, ICC'83*, Boston, U.S.A. pp. 375-379.

Beaupre, D.M., Le-Ngoc, T. and Zavitz, H.J. (1984) A new point-to-multipoint microwave radio system, *6th International Symposium on Subscriber Loops and Services, ISSLS'84*, Nice, France, (1) pp. 47-51.

Bellcore (1989) Generic requirements for Basic Exchange Radio Systems *Technical Advisory TA-TSY-000911* Issue 2, July 1989. Further details from Bellcore Document Registrar, 445 South Street, P.O. Box 1910, Morristown, New Jersey, 07960-1910, U.S.A.

Berceli, T., Frigyes, O.I. and Varady-Szabo, M. (1984) Some development results in rural radio systems, *IEEE International Conference on Communications, ICC'84*, Amsterdam, Netherlands, (2) pp. 965-9.

Berndt, A. (1986) RURTEL — A point-to-multipoint digital radio relay system for local subscriber links in rural areas, *European Conference on Radio Relay*, ECRR, Munich, Germany, pp. 268-275.

Blake, R.G. (1989) The role of radio for fixed local access, *Br Telecom Technol J.*, **7** (2) pp. 123-135.

Bonnerot, G., Floury, G. and Tanguy, R. (1987) IRT 2000 A versatile multipurpose radio system for subscribers, *IEEE/IECE Global Telecommunications Conference, Globcom'87*, Tokyo, Japan, pp. 169-173.

CCIR (1990a) *Recommendations of the CCIR, XVIIth Plenary Assembly*, Dusseldorf, 1986, Vol V — Propagation in non-ionized media. See also Annex to Vol V — Reports of the CCIR, 1990.

CCIR (1990b) *Recommendations of the CCIR, 1990, XVIIIth Plenary Assembly*, Dusseldorf, 1990, Vol IX, Part 1 — Fixed Service using radio-relay systems. See also Annex to Vol IX, Part 1 — Reprots of the CCIR, 1990.

CCIR (1987) Conclusions of the Interim Meeting of Study Group 9, Part 1 (Fixed service using radio relay systems) Geneva, Nov/Dec.

CCITT (1989a) Telephone Network and ISDN. Quality of Service, Network Management and Traffic Engineering, Blue Book Vol II — Fascicle II.3, *Recommendations E.401-E.880*. Geneva.

CCITT (1989b) General characteristics of International Telephone Connections and Circuits, Blue Book Vol III — Fascicle III.1 *Recommendations G.101-G.181*, Geneva.

CCITT/CCIR (1976) *GAS-3 Economic and technical aspects of the choice of transmission systems*, Geneva.

Chas, P.L. et al. (1986) A new generation of TDMA point to multipoint systems: Architecture and concepts, *European Conference on Radio Relay, ECRR*, Munich, West Germany, pp. 260-267.

Chas, P.L. and Jimenez, J. (1987) Applications and perspectives of ISDN multiaccess systems for rural and urban applications — a view from Spain, *IEEE/IECE Global Telecommunications Conference, Globcom'87*, Tokyo, Japan, pp. 174-177.

Cox, D.C. (1982) Co-channel interference considerations in frequency re-use small coverage area radio systems, *IEEE Trans. Comms*, **COM-30**, (1).

Daboual, H., Floury, G. and Garnier, C. (1981) IRT 1500 Rural telephony system, *Commutation and Transmission*, **3**, (3) pp. 39-53.

De Couesnongle, M. and Garnier, C. (1983) IRT 1500 Integrated rural telephony system, *Philips Telecommunications Review*, **41**, (2).

De Couesnongle, M., Floury, G. and Tanguy, R. (1987) IRT 2000 subscriber multiple access radio system, *Commutation and Transmission*, pp.15-18.

Dupuis, P. et al. (1983) Millimetre wave subscriber loops, *IEEE Journal on Selected Areas of Communications*, **SAC-1** (4) pp. 623-632.

Fichaut, J., et al. (1985) New services for the IRT 1500 subscriber connection system, *Communtation and Transmission*, **2**, pp. 25-38.

Forestier, A. (1985) UDS 10-64 — An urban data transmission system on TDMA microwave radio, *IEE Conference Publication No. 246*, pp. 97-100.

Garnier, C.R. and Chabert, J.L. (1981) The IRT 1500, a digital system serving rural zone subscribers by TDMA radio, *International Switching Symposium – ISS'81*, Montreal, Canada. (4) pp. 41C6/1-5.

Hampton, W.R. (1989) Applications and economics of the new exchange radios for basic exchange telecommunications radio service (BETR), *IEEE Global Telecommunications Conference, Globecom'89*, Dallas, U.S.A. pp.1045-8.

Hart, P. (1988) A point-to-multipoint digital radio system for rural subscriber areas, *IEE International Conference on Rural Telecommunications*, London, U.K., Conference publication No.286, pp. 80-83.

Hewitt, M.T., Scott, R.P. and Ballance, J.W. (1984) A cost effective 19GHz digital multipoint radio system for local distribution applications, *IEEE International Conference on Communications*, ICC'84, Amsterdam, Netherlands, pp. 31.4.

Hiyama, T. et al. (1985) Digital Radio Concentrator System *NEC Res. and Develop.*, (76) pp. 24-35.

Hiyama, T. et al. (1986) 1.5GHz transmitter-receiver for Digital Radio Concentrator System (DRCS), *NEC Res. and Develop.*, (81) pp. 86-91.

Holm, V. (1988) Point to multipoint microwave in spanning the globe, *Global Communications*, Second Quarter, pp. 20–23.

Hoppitt, C.E. and Clarke, D.E.A. (1989) The provision of telephony over passive optical networks, *Br Telecom Technol J.*, **7** (2) pp. 100-113.

James, S.W., Hooppell, A.R. and Mohamed, S.A. (1990) Point-to-multipoint radio in the British Telecom network, *Second IEE International Conference on Rural Telecommunications*, London, UK, Conference Publication no. 328, pp. 85–90.

Latorre, M.A., Tellado, A. and Chas, P.L. (1987) SMD 30/1'5: An advanced P.M.P. TDMA radio system with powerful operation and maintenance facilities, *IEEE International Conference on Communications, ICC'87*, Seattle, U.S.A. (1) pp. 10.3.1-5.

Lawson, I.C. and Ellershaw, J.C. (1986) Co-channel interference in a cellular rural radio telephone system, *IEEE International Conference on Communications, ICC'86*, Toronto, Canada, pp. 55.1.

Le-Ngoc, T. (1985) A random-request demand-assigned multiple access protocol for point-to-multipoint radio systems, *IEEE International Conference on Communications, ICC'85*, Chicago, U.S.A., pp.744-748.

Le-Ngoc, T. and Zavitz, H.J. (1985) A point-to-multipoint microwave radio system with a random-request demand-assignment time division multiple-access (RR-DA-TDMA) scheme, *IEE Conference Publication No. 246*, London, U.K. pp. 93-96.

Le-Ngoc, T. (1986) SR500 — A point-to-multipoint digital radio system, *IEEE International Conference on Communications, ICC'86*, Toronto, Canada, pp. 55.2.

Le-Ngoc, T. Stashin, M. and El Kateeb, A. (1989) ISDN signalling and packet data support on point-to-multipoint subscriber radio systems, *IEEE International Conference on Communications, ICC'89*, Boston, U.S.A. pp. 1314-1318.

Lin, S.H. and Wolff, R.S. (1990) Basic exchange radio — from concept to reality, *IEEE International Conference on Communications, ICC'90*, Atlanta, U.S.A. pp. 206.2.1-206.2.7.

Manichaikul, Y. Silverman, D. and Szeliga, J.J. (1983) RAPAC — a point-to-multipoint digital radio system for local distribution, *IEEE International Conference on Communications, ICC'83*, Boston, U.S.A., pp. 1013-16.

Marchand, P. (1986) 23GHz digital transmission system DTS 10-64 *European Communications on Radio Relay*, Munich, West Germany, pp. 284-290.

McGuire, R.J. (1989) Exchange radio technology *IEEE Global Telecommunications Conference, Globecom'89*, Dallas, U.S.A. pp. 29.A.4.

Mohamed, S.A. and Ballance, J.W. (1985) 19GHz digital point-to-multipoint radio system for local distribution, *IEEE International*

Conference on Communications, ICC'85, Chicago, U.S.A., pp. 735-739.

Moris, M.J. and Zavitz, H.J. (1988) A new high capacity TDMA radio system, *IEE International Conference on Rural Telecommunications, Conference Publication No. 286*, pp. 74-79.

Mullen, J.F. (1989) Wireless digital access, *IEEE Global Telecommunications Conference, Globecom'89*, Dallas, U.S.A. pp. 1036-1044.

Murakami, M. et al. (1983) A multiple access digital microwave radio system for local subscribers, *IEEE International Conference on Communications, ICC'83*, Boston, U.S.A., Vol.1, pp. 380-6.

Nakayama, H., Yoshida, T. and Tanaka, K. (1985) 26GHz band digital subscriber radio system (26SS-D1) for high speed digital communications, *IEEE International Conference on Communications, ICC'85*, Chicago, U.S.A., pp. 729-734.

Oakley, K.A., Taylor, C.G. and Stern, J.R. (1988) Passive fibre loop for telephony with broadband upgrade, *Proceedings of the International Symposium on Subscriber Loop and Services, ISSLS'88*, Boston, U.S.A. pp. 179-183.

Ogawa, H., Yamamoto, K. and Imai, N. (1984a) A 26GHz high performance MIC transmitter/receiver for digital radio subscriber systems, *IEEE Transactions on Microwave Theory and Techniques*, **MIT-32**, (12) pp. 1551-1555.

Ogawa, E. and Satoh, A. (1984b) Radio zone design using visibility estimation for local distribution systems in metropolitan areas, *IEEE International Conference on Communications, ICC'84*, Amsterdam, Netherlands, pp. 946-950.

Okubo, H., Miura, S. and Nagasawa, S. (1988) Building block design of large capacity PCM-TDMA subscriber system and direct digital interface to digital exchange, *IEE International Conference on Rural Telecommunications*, London, U.K., Conference Publication No. 286, pp. 69-73.

Omtveit, O. (1989) Development of a radio network, *Communications Engineering International*, **10**, (10), pp. 46-48.

Raju, G.S. and Prasad, K.V.K.K. (1990) A TDMA point to multipoint rural radio system for trunking and local loop applications, *Sec-*

ond *International Conference on Rural Telecommunications*, London, U.K., Conference Publication No. 328, pp. 91-5.

Rhese, J.K. (1986) The logical first/last mile digital termination system, *IEEE International Conference on Communications, Icc'86*, Toronto, Canada.

Richman, G.D. Chisholm, J.A. and Smith, P.C. (1990) Transmission of synchronous digital hierarchy signals by radio, *IEEE International Conference on Communications, SuperComm/ICC'90*, Atlanta, U.S.A.

Sasaki, S. et al. (1983) 2GHz multi-direction time division multiplex radio equipment, *IEEE International Conference on Communications, ICC'83*, Boston, U.S.A. (1) B2.2.1-5.

Satellite Communications (1990) Motorola plans to marry satellite and cellular; Inks deals with Inmarsat, AMSC and TMI, August, pp.11.

Saunders, R.G. (1989) Ultraphone — wireless digital loop carrier system, *Proceeding of National Communications Forum*, **42**, (2) pp. 1860-6.

Scott, R.P. (1984) A low cost 19GHz radio sub-system for point-to-multipoint radio applications, *Br Telecom Technol J.* **2**, (3) pp. 50-57.

Shindo, S., Kurita, O. and Akaike, M. (1981) Radio subscriber loop system for high speed digital communications, *IEEE International Conference on Communications, ICC'81*, Denver, U.S.A., pp. 66.1.1-5.

Shindo, S. et al. (1983a) TDMA for local distribution system, *IEEE International Conference on Communications, ICC'83*, Boston, U.S.A., pp. 370-374.

Shindo, S. et al. (1983b) Radio local distribution system for high speed digital communications, *IEEE Journal on Selected Areas in Communications, SAC-1, (4), pp.609-615.*

Stern, M. (1987) Cellular technology — revisited for rural service, *Communications International*, Oct., pp. 96-100.

Takada, M. (1983) A multidirection time division multiplex system for point-to-multipoint communication, *NEC Res. and Develop.*, (71) pp. 9-19.

Yamada, T., Tajima, K. and Aikawa M. (1986) High capacity sub-scriber radio, *IEEE International Conference on Communications, ICC'86*, Toronto, Canada, **3** pp. 1758-1762.

Yoshida, T. (1984) Digital subscriber radio system in the 26GHz band, *Japan Telecom. Rev.*, **26**, (3), pp. 188-194.

7. Acronyms

Every discipline has its own 'language' and this is especially true of telecommunications, where acronyms abound. In this guide to acronyms, where the letters within an acronym can have slightly different interpretations, these are given within the same entry. If the acronym stands for completely different terms then these are listed separately.

ADSL	Asymmetrical Digital Subscriber Loop. (Technique for providing broadband over copper.)
AGC	Automatic Gain Control.
ANSI	American National Standards Institute.
ASCII	American Standard Code for Information Interchange. (Popular character code used for data communications and processing. Consists of seven bits, or eight bits with a parity bit added.)
AWG	American Wire Gauge.
B-ISDN	Broadband Integrated Services Digital Network.
BABT	British Approvals Board for Telecommunications.
BCD	Binary Coded Decimal. (An older character code set, in which numbers are represented by a four bit sequence.)
BER	Bit Error Ratio. (Also called Bit Error Rate. It is a measure of transmission quality. It is the number of bits received in error during a transmission, divided by the total number of bits transmitted in a specific interval.)
BERT	Bit Error Ratio Tester. (Equipment used for digital transmission testing.)
BETRS	Basic Exchange Telecommunications Radio Service.

BEXR	Basic Exchange Radio.
BSI	British Standards Institute.
BS	Base Station. (Used in mobile radio systems.)
BSS	Broadcast Satellite Service.
BTS	Base Transceiver Station. (Used in mobile radio based systems to provide the air interface to the customer.)
BUNI	Broadband User Network Interface.
CARS	Community Antenna Radio Service.
CATA	Community Antenna Television Association.
CATV	Community Antenna Television. (Also refered to as Cable Television.)
CCI	Co-channel Interference. (Interference between two subscribers, using the same channel but in different cells, of a cellular mobile radio system.)
CCIR	Comite Consultatif Internationale des Radiocommunications. (International Radio Consultative Committee. Former standards making body within the ITU and now part of its new Radiocummunication Sector.)
CCITT	Comite Consultatif Internationale de Telephonique et Telegraphique. (Consultative Committee for International Telephone and Telegraphy. Standards making body within the ITU, now forming part of the new Standardisation Sector.)
CEPT	Conference des administrations Europeenes des Postes et Telecommunications. (Conference of European Posts and Telecommunications administrations. Body representing European PTTs.)
CO	Central Office. (Usually refers a central switching or control centre belonging to a PTT.)
CoCom	Co-ordinating Committee on Multilateral Export Controls.
CODEC	COder-DECoder.
COMSAT	Communication Satellite Corporation.
CPE	Customer Premise Equipment.

CT	Cordless Telephony. (CT1 is first generation; CT2 is second generation, and CT3 is third generation.)
DECT	Digital European Cordless Telephony. (Or Digital European Cordless Telecommunications. ETSI standard, intended to be a replacement for CT2.)
DES	Data Encryption Standard. (Public standard encryption system from the American National Bureau of Standards.)
DFM	Dispersive Fade Margin. (Used in microwave system design.)
DFT	Discrete Fourier Transform.
DOV	Data Over Voice. (Technique for simultaneous transmission of voice and data over telephone lines. This is a less sophisticated technique than ISDN.)
DPCM	Differential Pulse Code Modulation.
DS-0	Digital Signal level 0. (Part of the US transmission hierarchy, transmitting at 64kbit/s. DS-1 transmits at 1.544Mbit/s, DS-2 at 6.312Mbit/s, etc.)
DSRR	Digital Short-Range Radio.
DTE	Data Terminal Equipment. (User end of network which connects to a DCE. Usually used in packet switched networks.)
DTH	Direct to Home. (Usually refers to satellite TV.)
DTI	Department of Trade and Industry.
DTMF	Dual Tone Multi-Frequency. (Telephone signalling system used with push button telephones.)
ECS	European Communication Satellite.
ECSA	Exchange Carriers Standards Association (USA).
EFTA	European Free Trade Association.
EHF	Extremely High Frequency. (Usually used to describe the portion of the electromagnetic spectrum in the range 30GHz to 300GHz.)
EIA	Electronic Industries Association. (Trade association in USA.)

EIRP	Equivalent Isotropically Radiated Power. (Or Effective Isotropically Radiated Power. Of an antenna.)
ELT	Emergency Locator Transmitter.
E-MAIL	Electronic Mail.
EMC	Electromagnetic Compatibility.
EMI	Electromagnetic Interference.
EPIRB	Emergency Position Indicating Radio Beacon.
ERP	Equipment Radiated Power. (Also referred to as Effective Radiated Power of an antenna.)
ESA	European Space Agency.
ESB	Emergency Service Bureau. (A centralised location to which all emergency calls, e.g. police, ambulance, fire brigade, are routed.)
ESN	Electronic Serial Number. (Usually refers to the personal identity number coded into mobile radio handsets.)
ETE	Exchange Terminating Equipment.
ETNO	European Telecommunications Network Operators. (Association of European public operators.)
ETSI	European Telecommunications Standards Institute.
FCC	Federal Communications Commission. (US authority, appointed by the President to regulate all interstate and international telecommunications.)
FDM	Frequency Division Multiplexing. (Signal multiplexing technique.)
FDMA	Frequency Division Multiple Access. (Multiple access technique based on FDM.)
FEXT	Far End Crosstalk.
FFM	Flat Fade Margin. (Used in microwave system design.)
FHSS	Frequency Hopping Spread Spectrum.
FPLMTS	Future Public Land Mobile Telecommunication System. (ITU-T name for third generation land mobile system, now renamed IMT2000. See also UMTS.)

GAP	Groupe d'Analyse et de Prevision. (Analysis and Forecasting Group. A sub-committee of SOGT, part of the European Community.)
GATT	General Agreement on Tariffs and Trade.
GEO	Geostationary Earth Orbit. (For satellites. Approximately 36000km altitude.)
GEOS	Geodetic Earth Orbiting Satellite.
GPS	Global Positioning System. (Usually refers to satellite based vehicle positioning.)
GSM	Global System for Mobile communication. (Previously Groupe Special Mobile. Pan-European standard for mobile communication.)
GSO	Geostationary Satellite Orbit.
GVPN	Global Virtual Private Network.
HACBSS	Homestead & Community Broadcasting Satellite Services.
HCI	Human Computer Interface.
HD	Harmonisation Document. (Sometimes used to describe an EN.)
HF	High Frequency. (Radio signal.)
HRP	Horizontal Radiation Pattern. (Of an antenna.)
ICAO	International Civil Aviation Organisation.
IDA	Integrated Digital Access. (ISDN pilot service in the UK.)
IDN	Integrated Digital Network. (Usually refers to the digital public network which uses digital transmission and switching.)
IEC	International Electrotechnical Commission.
IEC	Interexchange Carrier. (US term for any telephone operator licensed to carry traffic between LATAs interstate or intrastate.)
IEEE	Institute for Electrical and Electronics Engineers. (USA professional organisation.)
IF	Intermediate Frequency.

IFL	International Frequency List. (List of frequency allocations published by the ITU.)
IFRB	International Frequency Registration Board. (Part of the ITU's Radiocommunication Sector.)
IM	Intermodulation.
IMO	International Maritime Organisation.
IMSI	International Mobile Subscriber Identity. (Personal number associated with a PCN user's handset, and issued on his SIM. It is the number which the network uses to identify the mobile.)
ISDN	Integrated Services Digital Network. (Technique for the simultaneous transmission of a range of services, such as voice, data and video, over telephone lines.)
ISO	International Standardisation Organization.
ISO	International Satellite Organisation. (Includes Intelsat, Inmarsta and Eutelsat.)
ITU	International Telecommunication Union.
ITU-T	International Telecommunication Union Telecommunication sector.
ITU-R	International Telecommunication Union Radiocommunication sector.
LAN	Local Area Network. (A network shared by communicating devices, usually on a relatively small geographical area. Many techniques are used to allow each device to obtain use of the network.)
LASER	Light Amplification by Stimulated Emission of Radiation. (Laser is also used to refer to a component.)
LEO	Low Earth Orbiting. (For satellites. Approximately 700km to 1500km altitude.)
LEOS	Low Earth Orbit System. (Satellite communication system with satellites not in geostationary orbit.)
LMSS	Land Mobile Satellite Services.
LOS	Line Of Sight. (Transmission system, e.g. microwave.)

LTE	Line Terminating Equipment. (Also called Line Terminal Equipment. Equipment which terminates a transmission line.)
MAN	Metropolitan Area Network.
MASER	Microwave Amplification by Simulated Emission of Radiation.
MATV	Mast Antenna Television. (Or Master Antenna Television. Local cable television system for a hotel or apartment block.)
MEO	Medium Earth Orbit. (For satellite. Approximately 10000km to 15000km altitude.)
METEOSAT	Meteorological Satellite.
MF	Multi Frequency. (A signalling system used with push-button telephones.)
MIFR	Master International Frequency Register. (Register of allocated international frequencies maintained by the IFRB.)
MIPS	Million Instructions Per Second. (Measure of a computer's processing speed.)
MMI	Man Machine Interface. (Another name for the human-computer interface or HCI.)
MMIC	Monolithic Microwave Integrated Circuit.
MoD	Ministry of Defence (UK).
MODEM	Modulator/Demodulator. Device for enabling digital data to be send over analogue lines.
MOS	Metal Oxide Semiconductor. (A semiconductor technology.)
MPG	Microwave Pulse Generator. (A device for generating electrical pulses at microwave frequencies.)
MPT	Ministry of Posts and Telecommunications (Japan).
MSC	Mobile Switching Centre. (Switching centre used in mobile radio systems.)
MSS	Mobile Satellite Service.
NEP	Noise Equivalent Power.
NF	Noise Figure.

NPR	Noise Power Ratio.
OEM	Original Equipment Manufacturer. Supplier who makes equipment for sale by a third party. The equipment is usually disguised by the third party with his own labels.)
OFTEL	Office of Telecommunications. (UK regulatory body.)
OS	Outstation.
OTS	Orbital Test Satellite.
PDH	Plesiochronous Digital Hierarchy. (Plesiochronous transmission standard.)
PDM	Pulse Duration Modulation. (Signal modulation technique, also known as Pulse Width Modulation or PWM.)
PEP	Peak Envelope Power.
PFD	Power Flux Density. (Measure of spectral emission strength.)
PLMN	Public Land Mobile Network.
POTS	Plain Old Telephone Service. (A term loosely applied to an ordinary voice telephone service.)
PPL	Phase Locked Loop. (Component used in frequency stability systems such as demodulators for frequency modulation.)
PSTN	Public Switched Telephone Network. (Term used to describe the public dial up voice telephone network, operated by a PTT.)
PTN	Public Telecommunications Network.
PTO	Public Telecommunication Operator. (A licensed telecommunication operator. Usually used to refer to a PTT.)
PTT	Postal, Telegraph and Telephone. (Usually refers to the telephone authority within a country, often a publicly owned body. The term is also loosely used to describe any large telecomunications carrier.)

QoS	Quality of Service. (Measure of service performance as perceived by the user.)
RA	Radiocommunications Agency. (UK body responsible for frequency allocation.)
RACE	Research and development in Advanced Communication technologies in Europe.
RARC	Regional Administrative Radio Conference.
RBER	Residual Bit Error Ratio. (Measure of transmission quality. ITU-T Rec. 594-1.)
RBOC	Regional Bell Operating Company. (US local carriers formed after the divestiture of AT&T.)
RCU	Remote Concentrator Unit.
RF	Radio Frequency. (Signal.)
RFI	Radio Frequency Interference.
RITL	Radio In The Loop. (Radio used for the subscriber's local loop.)
RPFD	Received Power Flux Density.
SDH	Synchronous Digital Hierarchy.
SHF	Super High Frequency.
SIM	Subscriber Identity Module. (Usually a plug in card used with a mobile radio handset.)
SIO	Scientific and Industrial Organisation.
SONET	Synchronous Optical Network. (Synchronous optical transmission system developed in North America, and which has been developed by ITU-T into SDH.)
SQNR	Signal Quantisation Noise Ratio.
TDRS	Tracking and Data Relay Satellite.
TE	Terminal Equipment.
TTE	Telecommunication Terminal Equipment.
UHF	Ultra High Frequency. (Radio frequency, extending from about 300MHz to 3GHz.)
UI	User Interface.

UL	Underwriters Laboratories. (Independent USA organisation involved in standards and certification.)
UMTS	Universal Mobile Telecommunications System. (ETSI terminology for a future public land mobile telecommunications system.)
UPT	Universal Personal Telecommunications. (ITU-T concept of the personal telephone number.)
USART	Universal Synchronous/Asynchronous Receiver/Transmitter. (A device, usually an integrated circuit, used in data communication devices, for conversion of data from parallel to serial form for transmission.)
VADS	Value Added Data Service.
VAN	Value Added Network
VANS	Value Added Network Services.
VAS	Value Added Service. (See also VANS.)
VASP	Value Added Service Provider.
VBR	Variable Bit Rate.
VDU	Visual Display Unit. (Usually a computer screen.)
VF	Voice Frequency. (Signalling method using frequencies within speech band. Also called in-band signalling. Also refers to the voice frequency band from 300Hz to 3400Hz.)
VFCT	Voice Frequency Carrier Telegraph.
VHF	Very High Frequency. (Radio frequency in the range of about 30MHz and 300MHz.)
VLF	Very Low Frequency. (Radio frequency in the range of about 3kHz to 30kHz.)
WAN	Wide Area Network.
WARC	World Administrative Radio Conference. (Of ITU. Now known as WRC.)
WATTC	World Administrative Telephone and Telegraph Conference. (Of ITU.)
WRC	World Radio Conference. (Of ITU.)

Index